J

Chem-Bio Handbook

Second Edition

UK and International English
Language Version

Adapted by

Adrian Dwyer

John Eldridge

Mick Kernan

Jane's Information Group

Editorial/Project Management
Coordinating Editor: *Robert Fanney*

Editorial
Content Developer: *Rennie Campbell*
Content Developer: *Jim Tinsley*
Chief Content Development Officer: *John Boatman*

Production
Content Services Director: *Anita Slade*
Global Production Services Manager: *Jane Lawrence*
Content Editing Manager: *Jo Agius*
Production Editor: *Neil Grace*

Front Cover Design
Senior Design Manager: *Steve Allen*
Coordinating Editor: *Rob Fanney*

Administration
Content Development Manager: *Ian Synge*
Chief Content Development Officer: *John Boatman*
Chief Executive Officer: *Alfred Rolington*

1st printing September 2002
2nd printing April 2003

"Jane's" is a registered trademark

Published by Jane's Information Group
Sentinel House, 163 Brighton Road, Coulsdon,
Surrey CR5 2YH, UK
Tel: (+44 20) 87 00 37 00; Fax: (+44 20) 87 00 10 06
e-mail: info@janes.co.uk

Printed in the UK

British Library Cataloging-in-Publication Data

**A catalogue record of this book is available from the
British Library**

ISBN 0-7106-2571-5

Publisher's Note
Based on *Jane's Second Edition Chem-bio Handbook* the UK and international English language version leverages knowledge and expertise in the UK emergency response community to provide this substantially updated resource. Essential to this extensive redesign process were:

Adrian Dwyer, lead editor and writer of *Jane's Chem-bio Handbook International* has over 15 years of experience dealing with CBRN response issues, firstly as an Army bomb disposal officer and high-risk search adviser and secondly as the counter-terrorism advisor to the British Transport Police. Adrian is a member of the Risk and Security Management Forum and the Institute of Explosives Engineers. He holds an MSc in risk, crisis and disaster management and contributed to the 1999, Jane's Chemical – Biological Defense Guidebook.

John Eldridge provided invaluable feedback and assistance in reworking current content. A Commander with 25 years of experience in the Royal Navy he has served as Naval Staff expert on toxic defence, survivability and shipping safety. During Operation Desert Storm, he served as Principal Advisor to the MOD's Assistant Chief of General Staff on NBCD preparation and management. Since 1997, he has served as Editor of Jane's Nuclear, Biological and Chemical Defence.

Mick Kernan provided extensive assistance during the rework and was especially helpful in understanding Chem-bio related HAZMAT response issues. Mr. Kernan has been active in the fire services at various levels for over 32 years. A decorated fireman, he has served as Staff Officer to four Commandants at the Fire Services College in Moreton in Marsh and is currently serving as Historian and Archivist.

This handbook is based on original work by authors Dr. Ken Alibek, Dr Frederick Sidell, Dr. William Patrick, Thomas R. Dashiell and Dr. Scott Layne, and on editorial review by:

Lt. Col. Robert C. Allen Chief of Aeromedical Evacuation US Air Force School of Aerospace Medicine

Lt. Col. Robert L. Domenici Commander 2nd Civil Support Team (Weapons of Mass Destruction) New York National Guard.

Colonel Edward Eitzen, MD Chief of Operational Medicine Division, USAMRIID

Colonel Charles Hurst, MD Chief of Chemical Casualty Care Division, USAMRICD

Anne Keleher Research Scientist, biological threat agents, Advanced Biosystems Inc

Dr. Mark Keim Co-Lead, Emergency Operations Center, Centers for Disease Control

Sergeant John Sullivan Planning Coordinator, Special Threats, Los Angeles County Sheriff's Department

Major John Woods, MD Chief Hospital, Installation and Disaster Preparedness, Operational Medicine Division, USAMRIID

Gary Briese, Executive Director, International Association of Fire Chiefs

Major Katie Carr, US Army Medical Research & Material Command

Lt. Col. Ciesiak, MD Operational Medical Division, USAMRIID

Dr. Scott R. Lillibridge, Centers for Disease Control

Robert D. Paulison, Fire Chief, Miami Dade Fire Rescue

Major Julie Pavlin, MD, Operational Medicine Division USAMRIID

Col. Rasa Silenas, Chief, Disaster Preparedness, 59th Medical Wing, US Air Force

Special thanks must also be given to Marleen Wong, Director of Mental Health Services, District Crisis Teams and to Michael Hopmeier of Unconventional Concepts, Inc (UCI) who was instrumental in assembling the final editorial panel of experts which reviewed and commented on the manuscript and made many important additions.

TABLE OF CONTENTS

CHAPTER 1: PRE-INCIDENT PLANNING

CHAPTER 2: ON-SCENE PROCEDURES

CHAPTER 3: CHEMICAL AGENTS

CHAPTER 4: CHEMICAL TREATMENT

CHAPTER 6: BIOLOGICAL TREATMENT

CHAPTER 7: POST INCIDENT

CHAPTER 8: APPENDIX

CHAPTER 1: PRE-INCIDENT PLANNING

CHAPTER 1: PRE-INCIDENT PLANNING

Chemical or Biological (C or B) incidents associated with terrorism are frequently characterised as 'worst case scenarios', against which there are few defences. While it is certainly possible that an attack could be catastrophic to unprotected personnel, defences and countermeasures do exist. It would be misleading to characterise all C or B threats as representing the same serious risks as those once posed by the *battlefield* ordnance of the Former Soviet Union (FSU).

In a no-notice scenario it is clear that responders will need to be alert to the earliest signs or symptoms of C or B attack. In real terms the chemical scenario should be easier to identify because of the rapid onset of symptoms among those affected. A biological incident will represent a more complex scenario. Unless supported by a warning, intelligence or the discovery (and recognition) of a dispersal device, the first indication that an attack has taken place may be the ensuing 'public health' emergency. In relation to many biological agents, this may not occur until several days after exposure.

Responders with specialised training will be best placed to:

◆ Identify, assess and report on-site hazards
◆ Assist casualties and evacuees
◆ Contain the incident
◆ Carry out on-site activities in accordance with the national strategy and/or local policies
◆ Manage the return to normality

However, during the early stages of an incident, the risks will be difficult to quantify and the hazards are potentially great for all responders, protected or not.

This handbook has been prepared specifically with first responders in mind. It does not supersede official guidance or operating procedures and its purpose is to inform the decision making process of first responders and provide an accessible source of useful information. The term *first responder* is used throughout the handbook. It is a generic description and is most applicable to members of the various agencies that will be involved in the immediate response to a suspected chemical or biological incident. The term does not imply a set level of training or equipment.

1.1 The assessment process

1.1.1 Threat assessment
The designated threat level will provide an initial starting point when determining the likelihood of an attack.

However, it should not be assumed that intelligence of this type will always be (a) available and (b) specific to a time, a place, a perpetrator or a method of attack. In relation to the hazards posed by C or B materials, as opposed to explosive hazards, it is reasonable to assume that people represent the most likely target. Pre-incident planning may therefore involve a subjective analysis of where people are most vulnerable.

1.1.2 Vulnerability assessment

In some cases, it may be possible to identify a particular target or target group. For example, the judiciary was targeted by Aum Shinrikyo (ASK) in 1994 and political figures were subject to anthrax attacks in the US and hoaxes in the UK in 2001. In this type of scenario a clear focus for pre-incident planning can be established. However, if the population at large is considered vulnerable, it will be necessary to examine a broader range of issues. For example:

◆ Where do large groups of people regularly congregate? (for instance high-rise offices; transportation centres; shopping centres; places of entertainment, worship or education are a few examples)
◆ What is the status of local security arrangements? (access control, CCTV and so on)
◆ Which locations could be considered confined spaces?

◆ Which locations are reliant on air-conditioning systems?
◆ Do some locations have a particularly high profile? (cultural, symbolic, media-related and so on)
◆ Is a particular location of economic significance? (power supplies, for example)
◆ Is the location associated with VIP events?

In reality, a target that is vulnerable to an explosive attack is probably equally vulnerable to a C or B attack. However, the issues listed above may be of greater relevance in a C or B scenario because they extend the scope of terrorist options.

1.1.3 Risk assessment
In assessing the likelihood of a C or B event and in determining a measure of the likely consequences, it is important to remember the overarching relevance of public health issues. In general terms, a chemical attack (identified by the rapid onset of symptoms) will typically involve the immediate (and possibly unco-ordinated) movement of unprotected persons from the scene. The release of a biological material, however, may necessitate *immediate* evacuation on a smaller scale and in a more co-ordinated manner. (This is because of the risk associated with the spread of contamination as people leave the scene). In either eventuality, it is important that casualties and those carrying contamination on their clothing receive the most appropriate help as soon as possible.

In essence, the alternatives can be characterised as 'keep and treat' or 'scoop and run.' To optimise the first responder capability, it is important to have established which option is to be employed in which scenario. In many cases, decontamination and treatment adjacent to the scene is likely to be the preferred option. 'Keep and treat' minimises the potentially uncontrolled spread of contamination and provides a focus for the medical response.

In no-notice attack scenarios, when casualties are likely to leave the scene before a police cordon is established, it will be necessary to implement special arrangements at local hospitals:

◆ Accident and Emergency (A and E) departments will have to be identified and notified
◆ A means of identifying and segregating contaminated casualties will be required
◆ Decontamination facilities should be established (or, where already in place, activated). Remember, decontamination arrangements should take into consideration vehicles and equipment as well as people.
◆ In relation to chemical materials, detection equipment (see section 8.1) will be required to establish the effectiveness of the decontamination process

1.2 Response planning

Wherever possible, tactical and strategic protocols for dealing with C or B incidents will have been devised in accordance with existing major incident procedures. The Fire Service, because of its HAZMAT experience, will be able to provide advice in relation to chemical hazards, decontamination considerations and cordon requirements. However, the incident location itself is a crime scene and will be managed accordingly (see Chapter 2). In the interests of effective multi-agency interaction, it is important that all responders understand not only their own responsibilities, but how the scene will be managed. For example:

◆ Special access control procedures (for vehicles and responders on foot):
  • **How will the site be accessed? How will primary and secondary routes be identified and notified?**
◆ Cordon management:
  • **Where will sentries with detection, identification or monitoring equipment be located?**
◆ **What are the actions to be taken if the cordon is breached (from within or without)?**
◆ Decontamination protocols:
  • **How will contaminated waste be managed?**
    – for example, water run-off, discarded clothing and so on
◆ Monitoring procedures

◆ Site evacuation arrangements
◆ Maintenance of an **incident log**

1.3 Equipment considerations

1.3.1 Personal Protective Equipment (PPE)

The level of protection afforded by PPE will limit the role of first responders at a suspected or confirmed scene (see Appendix B re Levels of Protection). Particularly in scenarios where the causative agent is unknown, air-purifying respirators (as opposed to Self-Contained Breathing Apparatus (SCBA)) will not necessarily provide adequate protection for those required to enter the scene.

First responders with PPE should be:
◆ Trained to the appropriate standard
◆ Fully aware of their role and the limitations of their PPE
◆ Competent in the use of PPE (including canister changing drills, if appropriate)
◆ Responsible for the maintenance of their own equipment and able to perform *buddy-buddy* checks with their colleagues
◆ Fully aware of any special storage issues/time-expiry requirements/safe disposal instructions

1.3.2 Decontamination equipment

Decontamination options designed for battlefield use (and primarily concerned with the removal of liquid materials) are not universally applicable to scenarios associated with urban terrorism. However, if liquid contamination is suspected, the following options may be appropriate for use by first responders:

◆ Fuller's earth (supplied loose or in military DKP 1 or DKP 2 packs)
◆ Garden-type sprays containing a weak hypochlorite solution (nine parts water one part bleach)
◆ Wooden spatulas for the mechanical removal of a contaminant
◆ Plastic sheeting on to which contaminated clothing/equipment may be 'dropped'
◆ Plastic bags (ideally nylon) for the disposal of contaminated material/equipment. Note: as such items may need to be double or triple bagged, a good supply of bags may be required

When such material is issued, instructions governing its use (particularly its safe disposal) must also be considered. Immediate advice in this regard is available via the Fire Service's HAZMAT specialists and/or other specialist agencies.

1.3.3 Detection and monitoring equipment

1.3.3.1 Chemical materials
Chemical agent monitors are discussed in more detail in Appendix A. Whatever the system used, equipment should be:

◆ Maintained to the appropriate schedule (*Note:* CAM and similar equipment benefit from regular use)

◆ Incorporated in first responder continuation training to ensure operator familiarity
◆ Deployed with sufficient consumables (batteries, filters and so on) to support an extended deployment
◆ Used at possible 'target' locations during low- or no-threat scenarios to establish background readings (that is, used for profiling)

1.3.3.2 Biological materials

In terms of detecting biological materials, there are no similar detection systems currently available to first responders. However, there is an ongoing effort to provide the type of field assay test kits used successfully by some agencies in the US. This equipment cannot detect aerosolised material but has proved effective in the process of assessing 'white powder' incidents.

1.3.4 Ancillary equipment

The precise requirement for equipment will vary in relation to incident scenario and first responder role. The following, if not held already, may be useful:

◆ A compass
◆ A means of measuring windspeed
◆ A means of measuring temperature
◆ Binoculars/spotting scope (compatible with respirators)
◆ Stroboscopic markers
◆ Air horns
◆ Fluorescent spray paint
◆ Coloured marker tape

◆ Chalk (for marking military style NBC suits and so on)
◆ Strong adhesive tape for sealing plastic bags, doors, windows and so on

1.4 Call-out procedures

When devising call-out procedures, consideration should be given to the following:

◆ Who maintains the call-out list? Is it up to date?
◆ How often is the procedure tested?
◆ Do first responders hold equipment at home or does it have to be collected from a central location? (What happens if access is denied to the equipment store location?)
◆ Is there redundancy within the call-out system? (What is the fall-back position?)
◆ Do first responder agencies share details of their call-out systems with each other?

1.5 Training and exercises

Where standards for training are set centrally, it is important that first responders meet at least the minimum requirements. Levels of expertise may then be enhanced through continuation training and practical exercises. It should also be remembered that skills will degrade over time if not practised regularly. The core skills required to maintain the basic levels of proficiency are likely to include:

◆ Recognition of the signs and symptoms of chemical or biological incidents (*How might casualties present themselves? The significance of chemical odours, the likely appearance of materials, the experiences of others, and so on.*)

◆ An awareness of the likely hazards and countermeasures
◆ The correct wearing of PPE (*fault finding and so on*)
◆ The correct use of equipment (*setting-up procedures, fault finding, actions on a reading*)
◆ Emergency procedures (*actions on: 'a find', a casualty, rapid evacuation of the scene, and so on*)
◆ Decontamination considerations (*dry decontamination, wet decontamination, expectations of other agencies and so on*)
◆ Interaction with other agencies (*including the military, government agencies, scene of crime officers and so on*)
◆ Cordon procedures (*including public order issues, for example, keep people out? Keep people in?*)
◆ Shelter or evacuation considerations
◆ Re-occupation searches

While many of the subject areas listed above may be addressed through lectures or presentations, there is no substitute for practical exercises (see Appendix E for an example of a phased training guide).

CHAPTER 2: ON-SCENE PROCEDURES

CHAPTER 2: ON-SCENE PROCEDURES

On-scene procedures will be dependent on a number of variables including:
◆ Specific mandatory actions (including safe systems of work)
◆ The availability of resources
◆ An appreciation of the relevant hazards
◆ The type of location affected
◆ The nature of the threat
◆ The requirements and responsibilities of other agencies

The information set out in this chapter therefore seeks to provide a generic guide to assist first responders.

2.1 Initial response procedures

2.1.1 General considerations
The following actions should be considered when initial reports suggest a C or B incident has occurred (because of (a) casualties, or (b) the presence of other C or B indicators with or without casualties):
◆ Attempt to gain information from the scene by remote means (CCTV, telephone/radio contact/ aerial surveillance)
◆ Approach from what is determined to be a safe direction (upwind if possible). Continue to monitor your surroundings for signs or symptoms of terrorist activity/contamination.

◆ Consider the possibility of secondary devices (conventional or C or B)
◆ Establish if the perpetrators are at the scene (is EOD/armed support available?)
◆ Can the spread of contamination be established?
◆ What are the initial symptoms of any casualties?
◆ Is the response likely to encounter a Public Order situation? Remember, possible victims may be in a state of distress and may not act rationally.
◆ If a release is suspected, don PPE immediately
◆ Remember: the sight of responders in PPE may cause concern or agitation among some sections of the community

2.1.2 A suspected chemical release

A chemical release may be apparent because of the sudden appearance of casualties (without an apparent cause) or the sudden death or incapacitation of birds, rodents, domestic pets insects and so on. In this scenario, first responders should consider the following:

◆ The most appropriate level of protection for those entering the scene is Level A or Level B (see Appendix B). For this reason, the rescue of casualties remains the responsibility of suitably equipped officers – usually from the Fire Service.
◆ Responders with Level C protection may not be able to operate effectively within the inner cordon when a contact hazard or vapour contamination is suspected.

◆ It is likely that *Dress States* will vary (for example, full PPE where the risks are high, outer garments less respirator where the risk is assessed to be lower). All responders using respirators should be familiar with the signal to 'mask-up'.

◆ The inner cordon is likely to be **at least** 100 m from the incident scene or the extent of a suspected hazard. Typically, sentries with detection, identification or monitoring equipment will be positioned at the cardinal points of the compass. (Remember: not every chemical material likely to be used by terrorists will be readily detectable – see section 8.1).

◆ In addition to CAM sentries, all responders should be alert to other signs and symptoms of a release (people in a state of distress, the death of birds, other mammals or insects). It is unlikely that signs of airborne contamination will be visible. In rare cases, contaminants will be visible in the form of a cloud or mist.

◆ Unprotected responders (that is, those without immediate access to appropriate PPE) are unlikely to operate in the area adjacent to the inner cordon. This is because of the transient nature of the vapour hazard and hazards associated with the movement of potentially contaminated personnel.

◆ It is likely that cordon distances will be modified as the incident progresses. In particular, officers downwind are likely to be moved further away from the scene.

◆ It is possible that if many people have been affected, some will make their own way to the hospital. If this is the case, separate decontamination arrangements will be required and hospitals will need to be forewarned.

◆ If the decision is taken to treat people on-site ('keep and treat'), a suitable area will be needed upwind of the incident, but away from unprotected personnel.

◆ Not all chemical materials produce chronic symptoms instantaneously. Therefore, all potential casualties must be seen by trained medical personnel. The Ambulance Service is responsible for triage on-site.

◆ Throughout the incident, first responders should be alert to the possibility of further attacks or the rapid onset of illness among their colleagues or other persons present.

◆ In relation to crime scene considerations, every effort should be made to identify potential witnesses or perpetrators from among the casualties.

◆ Evidence should be treated in accordance with normal evidence handling procedures, bearing in mind the possibility of contamination and the wishes of the Senior Investigating Officer (SIO).

2.2 Chemical agent indicator matrix
Only those in appropriate PPE should approach victims. All others must maintain a safe distance. If you cannot determine whether an indicator is present, note your suspicions or relevant observations and continue.

Information on the presence/absence of an indicator should be gathered from all on-scene personnel/ victims.

To use Agent Indicator Matrix

(1) Put a tick in each row in which an indicator is present. Grey boxes signify indicators that are not applicable to a given agent.
(2) At the bottom of each page total the number of ticks in each column.
(3) Total all ticks from each page. The column with the highest percentage of indicators should be considered the agent most likely to be present.

The Indicator Matrix is based on data provided by the US *Defense Protective Service*, Pentagon. It is designed to give the best approximation of the agent used but is not to be considered definitive until confirmed by an appropriately qualified agency.

Key:

A = Nerve Agents

B = Blister Agents

C = Cyanide

D = Pulmonary Agents

E = Riot Control Agents

= Not Applicable	A	B	C	D	E
Appearance					
Prostration		■		■	■
Twitching		■		■	■
Convulsions		■		■	■
Coma		■		■	■
Bleeding from mouth		■			
Coughing		■		■	
Sneezing		■		■	
Vomiting		■		■	
Fasciculations (muscle contractions)		■	■	■	
Skin					
Cyanosis (skin is blue or purple)		■		■	
Grey area of dead skin	■		■	■	■
Pain, irritation	■				
Clammy		■		■	
Sweating, localised or generalised		■	■	■	
Eyes					
Small pupils		■	■	■	
Normal, large pupils					
Involuntary closing	■	■		■	
Tearing		■			

■ = Not Applicable	A	B	C	D	E
Eyes *(continued)*					
Burning, irritation	■				
Headache, pain around eye		■	■	■	■
Dim vision		■	■	■	■
Blurred vision		■	■	■	■
Burning pain in eyes	■				
Redness		■	■	■	
Respiratory					
Coughing		■			
Runny nose					
Tight chest (shortness of breath)		■			
Burning, irritation in nose	■				
Cardiovascular					
Slow heart rate		■	■	■	■
Fast heart rate		■			
Digestive system					
Defecation		■		■	
Nausea				■	
TOTAL					
Total Indicators	26	8	23	11	16

2.3 A suspected biological release

A biological release is unlikely to be encountered as an immediately apparent incident scene. This is because of the relatively long period between exposure and the onset of recognisable symptoms. It is also likely that in circumstances where a potential biological hazard is discovered, it will be some time before the material is positively identified.

2.3.1 Actions when a biological material is suspected but casualties are not present

Prior to arrival. Officers dealing with reports of possible biological materials should seek to:

◆ Establish contact with the informant by telephone
◆ Confirm there are no casualties (if casualties are present, refer to the advice for Chemical Incidents)
◆ Reassure the informant that specialist assistance is on the way. (Provide advance warning that officers will be wearing PPE as their unexpected arrival could cause unnecessary panic)
◆ Ensure the informant and anyone in the immediate area (for example, same room, same office) moves away from the potential hazard, but stays at the scene
◆ Ensure windows are closed and air conditioning is switched off
◆ Obtain the information listed below:
  (1) Full details of informant. Including:
     (a) Name

(b) Status (employee, member of the public, security officer, and so on)

(c) Where they are calling from

(d) How they can be contacted

(2) Full description of item or substance. Including:

(a) Is it a powder, liquid or solid?

(b) What colour is it?

(c) How much is there? (for example, a pinch, a tea-spoon, a cup full and so on)

(d) Does it look like anything they can recognise? (for example: salt, washing powder, oil, milk, coffee)

(e) Is the material affected by the normal movement of air? (that is, is it very fine like cigarette ash? Or is it heavy, like sand?)

(3) The exact location (is the location itself of significance?)

(4) What makes the item/substance suspicious?

(5) Is discovery of the item/substance linked to any suspicious activity?

(6) Have there been any related threats or warnings?

While this information is being collated, officers should don the appropriate level of protective equipment. Unless specified otherwise, this should not be lower than Level C (see section 8.2).

2.4 Arrival at the scene

2.4.1 Biological release

A suspicious material may be considered to pose a biological hazard because of its appearance (see Chapter 5), the scenario follows the pattern of previous confirmed incidents, specific intelligence or other credible information has been received. When this is the case, the location should be secured and an inner and outer cordon established in accordance with national minimum standards. If the release is believed to be contained within a building and casualties are not immediately apparent, the initial cordon may be restricted to building entrances and exits. At a potential B incident, the likely priority will be to identify those requiring decontamination and medical treatment before evacuation of the building is initiated.

2.4.2 Chemical release

It is probable that casualties will be apparent at the scene of a chemical release. Depending upon the persistence of the material and its lethality, the cordon at the scene is likely to be relatively large, typically at least 100 m, unless circumstances dictate otherwise. At a potential C incident, the likely priority will be to initiate external evacuation as soon as possible. The minimum evacuation distance is unlikely to be less than 100 m from the extent of known or suspected contamination.

2.5 The incident site

The diagram below represents one possible option for a site layout and illustrates the relationship between **'Hot'**, **'Warm'** and **'Cold'** zones (see Appendix C). Precise locations and distances will be based on local or national protocols and may vary in relation to factors such as wind direction, the toxicity of the material involved, and so on. When establishing the site, arrangments should be made at the earliest opportunity to conduct a reconnaissance or alternative locations in case the wind direction alters or other sources of contamination are identified.

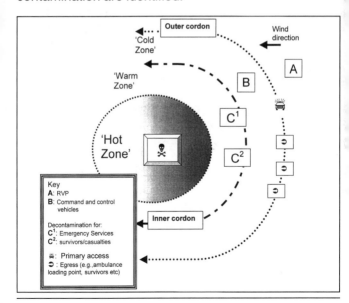

Key
A: RVP
B: Command and control vehicles

Decontamination for:
C¹: Emergency Services
C²: survivors/casualties

🚑: Primary access
↻: Egress (e.g.,ambulance loading point, survivors etc)

2.5.1 Responder actions

◆ Any persons who may have been exposed to biological material should be subject to dry decontamination procedures (the removal and bagging of outer garments) immediately. These items should be sealed in plastic bags (clear plastic if possible). Only after this has been done should 'wet' options be considered.

◆ Chemical casualties (confirmed or suspected) should be moved away from the vapour hazard as soon as possible

◆ Specialist advice on decontamination and immediate medical treatment should be sought from the Ambulance Service and health/medical advisers

◆ Formal documentation should be prepared concerning potential casualties and the movement of evacuees

◆ The scene should be marked in accordance with standard EOD procedures. For incidents inside buildings, doors or windows should not be wedged open.

◆ The crime scene should be managed in accordance with the instructions of the SIO

◆ People at the scene who have not been exposed to contamination should have their details recorded and, unless local procedures dictate otherwise, allowed to leave the scene

◆ First responders should remain vigilant for any indications of illness among colleagues or previously unaffected personnel (including evacuees)

2.6 Decontamination considerations

Decontamination will invariably involve the removal of items of clothing, which is known as dry decontamination. Consideration should always be given to (a) the victim's modesty and (b) some form of disposable 'coverall' and means of keeping the decontaminated person warm.

◆ In some circumstances it will also be necessary to apply decontamination liquids. Typically, bleach-type solutions are used for items of equipment and non-permeable materials. Soap and water are used increasingly to decontaminate people when the caustic effect of bleach is considered unacceptable. (Remember: bleach gives off chlorine gas and may be unsuitable for use in confined spaces)

◆ It should also be remembered that discarded clothing and cleaning materials are potential sources of contamination. The management of such waste should be carefully co-ordinated. In the short term, this may involve nothing more than placing the material within double plastic bags which are then sealed and placed in a safe location.

◆ If water is used during the decontamination process, it should be collected and treated in accordance with local procedures. However, if this is not possible, it should be diluted in accordance with advice from specialist agencies such as the fire service and related government departments (for example, government agencies responsible for public health and the environment).

◆ All items removed from people or used in the decontamination process should be treated in the first instance as hazardous waste.

The following outline-information for first responders is supplemented by more detailed advice in Appendix D.

2.6.1 Unprotected personnel

◆ **Clothing:** outer garments, including footwear should be removed as soon as possible. If contamination is suspected on underclothes, these should also be removed.

◆ **Skin:** once contaminated clothing has been removed, supervised decontamination with soap and water (ideally a low-pressure shower) should be initiated. (The purpose of supervision is to ensure the thoroughness of the washing process.)

2.6.2 Protected personnel

(*Note*: separate arrangements should already be in place for responders using gas-tight suits and SCBA)

PPE: outer garments should be removed and bagged. The respirator/breathing apparatus should not be removed at this stage. If PPE cannot be removed immediately, visible contamination may be removed by:

◆ Scraping with a stick
◆ The application of Fuller's earth (or similar absorbent materials)

If PPE is of the non-permeable type (Tyvek® for example) initial decontamination may be by high-pressure hoses or other types of field showering equipment. (*Note:* high-pressure systems are not suitable for exposed skin.)

After decontamination, all personnel should be checked for residual vapour using suitable detection, identification and monitoring equipment.

◆ **Other clothing:** if necessary, other items of clothing should be removed and bagged. The respirator/breathing apparatus should not be removed at this stage.
◆ **Skin:** showering with soap and water may be initiated before the respirator is removed

Once decontaminated, personnel should be escorted to a reception point where their welfare needs will be addressed and relevant details recorded.

2.7 Site set-up arrangements
Generic multi-agency arrangements are already in place in relation to set-up procedures for an incident scene. In relation to a C or B incident, the same general

considerations will apply. However, the following factors are particularly relevant.

◆ **Site boundaries:** inner and outer cordons are both likely to be modified in the following circumstances:
  • Changes in wind direction or windspeed
  • Movement of contaminated vehicles, personnel or material
  • Once the persistence/lethality of the hazard is confirmed
  • If a secondary device is suspected
  • On the advice of specialist agencies
  • If a subsequent attack is anticipated

Responders should be aware of the need to move quickly and of any pre-arranged signals indicating the need for immediate withdrawal (for example, air horns, maroons and so on).

◆ **Access control:** access control will be supervised in accordance with national arrangements. It is likely that any persons or vehicles leaving the outer cordon will be required to pass through a screening process.

◆ **Public order:** people at the scene of a suspected C or B incident (that is, within the inner or outer cordons) will fall under one of three categories:
  • **Uninjured/uncontaminated**
    Will be processed through the survivor reception centre. Their details must be recorded and they will usually be dealt with by follow-up agencies.

- **Injured**
  Are rescued, decontaminated and treated by the most appropriate emergency service. Injured persons must be documented before they leave the scene.
- **Dead**
  Should be left on scene unless their removal is essential as part of the rescue process.

◆ Remember: Any of these people could be the perpetrator. Injured and uninjured persons could be witnesses.

2.8 Management of evacuees

People who have not become incapacitated but are within the cordon may become agitated, hysterical or aggressive. They may ignore police advice and:

◆ Try to leave the scene before they are processed
◆ Attempt to remove PPE from first responders
◆ Attempt to harm themselves or others
◆ Refuse to follow police/emergency services' instructions

Keeping these people informed may alleviate some of their concerns. It is likely that all casualties and potential casualties will require reassurance that:

◆ The scene is under control
◆ Medical attention will be available more quickly if they follow instructions
◆ They will be evacuated from the scene as soon as it is safe to do so.

CHAPTER 3: CHEMICAL AGENTS

CHAPTER 3: CHEMICAL AGENTS

3.1 Chemical agent quick reference (continued overleaf)

	Nerve Agents (GA, GB, GD, GF, VX)	Cyanide (AC, CK)
Effects	Vapour: small pupils, runny nose, shortness of breath Liquid: sweating, vomiting Both: convulsions, cessation of respiration	Loss of consciousness, convulsions, temporary cessation of respiration
Onset	Vapour: seconds Liquid: minutes to hours	Seconds
First Aid	Atropine Diazepam	Amyl nitrite
Skin Decon	Soap and water	None usually needed
Detection	See section 8.1	

	Blister Agents (H, HD, L, CX)	Pulmonary Agents (CG, PFIB, HC)	Riot Control Agents (CS, CN)
Effects	Redness of the skin, blisters. Irritation of eyes. Cough, shortness of breath.	Shortness of breath, coughing	Burning, stinging of eyes, nose, airways, skin
Onset	Hours (immediate pain after Lewisite)	Hours	Seconds
First Aid	Immediate decontamination	None	None
Skin Decon	Soap and water	None usually needed	Water
Detection	See section 8.1		

3.2 General agent properties

Chemical Warfare (CW) agents have been thought of traditionally as weapons for use on the battlefield - to kill or injure an enemy. Indeed, chemicals contained within bombs, mines and projectiles have been used by armies around the world. In the First World War, for example, phosgene, chlorine, cyanide, riot control agents and sulphur mustard were all employed. The Iraq/Iran war and Iraqi attacks against its own population provide examples that are more recent.

On the battlefield, the CW agents still considered to be of greatest concern are sulphur mustard (a blister agent, or 'vesicant') and the various nerve agents. Both of these materials were held in great quantities by the former Soviet Union and some of its satellites. Available intelligence does not suggest that cyanide or the pulmonary agents (phosgene and chlorine) are likely to be encountered on the battlefield as 'weapons of choice'. Riot control agents were also used in the First World War and in Vietnam, but are considered to be law enforcement options, rather than combat weapons.

Particularly in relation to terrorism, chemical warfare agents are perceived increasingly as a threat in the civilian environment. The events in Matsumoto, Japan, (June 1994) and the Tokyo subway incident (March 1995) caused a major review of the risks associated with a no-notice chemical release in the urban environment. In relation to an unprotected civilian population, the release of chemical material is likely to cause casualties in significant numbers. **Cyanide and the nerve agents, for example, can cause death within minutes of exposure and immediate medical intervention is essential if lives are to be saved. Effective antidotes exist for both cyanide and nerve agents**.

Sulphur mustard and the pulmonary agents do not cause clinical effects until hours after exposure unless the amount of exposure is extremely high, and immediate intervention is rarely necessary. **There are no antidotes for mustard and pulmonary agents**.

Solid, Liquid or Gas
Chemical agents vary considerably in their properties and in their effects on humans. **Most CW agents are liquids**. The exception is riot control agents, which are solids that are dispersed as a fine powder or as an aerosol (a powder suspended in liquid).

Dissemination – Liquids may be disseminated by energy, such as heat to cause evaporation of the agent (as was done at Matsumoto), or by a force, such as in an exploding munition, or mechanical spray device. Alternatively, they may be allowed to evaporate spontaneously. After an explosive force is applied to a chemical material, part of the material will remain liquid, part will be aerosolised (droplets suspended in air), and part will evaporate to form a vapour. The aerosolised droplets will also evaporate over time - the rate of evaporation depends on a number of factors, including the ambient temperature, movement of air and relative humidity, for example.

Conversely, if no energy is applied to the liquid, it may evaporate slowly on its own, again depending on temperature and other factors. For example, the nerve agent placed on the Tokyo subways simply leaked from its container and evaporated spontaneously. The rate of evaporation was not high (about the rate at which water evaporates) and few people who were able to leave the scene quickly received a high concentration. (Further details are set out in the table below.)

CW agents that are gases under temperate conditions:	
Phosgene	44°F (7°C)
Cyanogen chloride	57°F (13.8°C)
Hydrogen cyanide	78°F (25.6°C)

Persistence

The volatility of chemical agents varies considerably, from the highly volatile pulmonary agent phosgene to the much less volatile nerve agent VX and vesicant mustard. **The less volatile the agent, the more persistent it is on the ground, in foliage or on vehicles.**

Persistence refers to the length of time an agent remains as a liquid. In the military, the terms 'persistent' and 'non-persistent' are used to characterise chemical agents. An agent is said to be 'persistent' if it remains as a liquid for longer than 24 hours and 'non-persistent' if it evaporates within that time.

Rate of evaporation – Many factors determine whether an agent will remain as a liquid or will evaporate. Foremost are the physical-chemical properties of the compound, but other factors include:

- Ambient temperature
- Windspeed
- Surface upon which the agent resides

An example is the difference between petrol and heavy motor oil. Under the same environmental conditions, petrol will evaporate faster - and thus be less persistent - than oil. Furthermore, on a hot day petrol will evaporate faster than it will on a cold day, and it will evaporate faster when placed in a strong breeze than it will in still air. Finally, it will evaporate faster from a polished floor than from a porous surface such as earth.

Chemical agents are no different from other liquids in this regard. For example, the nerve agent sarin, generally considered to be a non-persistent agent, will evaporate in about 2 hours from a sandy surface at 50°F (10°C) and in under 1 hour at 110°F (43°C). When placed on a chemical agent-resistant surface, sarin will evaporate in 15 minutes at 50°F (10°C) and in 2 minutes at 110°F (43°C). The vesicant mustard, generally considered to be a persistent agent, will evaporate from sand in about 100 hours at 50°F (10°C) and in about

7 hours at 110°F (43°C), and it will evaporate from a chemical-resistant surface in 12 hours at 50°F (10°C) and in 1 hour at 110°F (43°C) respectively.

Agent persistence: Most to least persistent:
VX
Tabun
Mustard
Lewisite
Sarin
Hydrogen cyanide
Cyanogen chloride
Phosgene
Chlorine

Evaporation time:
◆ **Sarin**
  • Sandy surface
  – 50°F (10°C) = 2 hours
  – 110°F (43°C) = 1 hour
  • Chemical-resistant surface
  – 50°F (10°C) = 15 minutes
  – 110°F (43°C) = 2 minutes
◆ **Mustard**
  • Sandy surface
  – 50°F (10°C) = 100 hours
  – 110°F (43°C) = 7 hours
  • Chemical-resistant surface
  – 50°F (10°C) = 12 hours
  – 110°F (43°C) = 1 hour

Effects

Some chemical agents cause serious clinical effects in humans almost immediately, while others cause minor or no clinical effects within the first minutes to hours after contact. For example:

◆ **Inhalation of a large concentration of a nerve agent or cyanide can cause loss of consciousness and convulsions within seconds**

◆ A high concentration of the pulmonary agent **phosgene may produce irritation in the eyes and nose initially,** but the **major effect of this agent will not appear until hours later**

◆ **Riot control agents produce** almost immediate **irritation and burning of the eyes, nose and upper airways,** but these rarely produce serious effects

◆ The **blister agent mustard causes no clinical effects for hours, but chemical changes in cells begin within minutes of contact**

Early recognition

A first responder entering an area in which there are many casualties from an unknown substance **must quickly evaluate the scope of the likely hazards.** A responder must **beware of contamination** from:

◆ Liquid that might be on the ground
◆ Adjacent objects
◆ Vapour that is produced when this liquid evaporates ('off-gassing') from objects including victims' clothing

The **diagnosis will be made by trained medical personnel from clinical signs and symptoms**. Chemical agent detectors and alarms are a useful early warning system, if available and if able to detect the material used.

First responders should appreciate that **no one casualty is likely to have all the 'classical' signs and symptoms of the chemical responsible**. The **full spectrum of effects over many casualties must be evaluated**.

The most severe effects within a short period of time are caused by **nerve agents and cyanide**. They can both **produce loss of consciousness, convulsions and death within minutes**. If any casualties have these effects, the responder should look for other effects of these agents in the less severe casualties.

◆ **Shortness of breath is an early effect of** several of these chemicals. The **riot control agents** and a high concentration of **phosgene** produce this by irritation of the airways, while **nerve agents** cause bronchoconstriction.

◆ Riot control agents and phosgene, along with **chlorine**, also irritate the eyes and nose, and **nerve agents are likely to cause miosis (small pupils)** and rhinorrhoea (runny nose), though without irritation

◆ **Burning or irritation of the eyes**, the nose, the upper airways and the skin **can be caused by** several of these agents

◆ With some agents - **riot control agents**, **pulmonary agents** and **cyanogen chloride** - these effects decrease with time

◆ In relation to the vesicants - **Lewisite** and **phosgene oxime** – the effects worsen over time because of tissue damage

More detailed information on early recognition of CW agent effects is provided under each of the following agent descriptions and casualty management is detailed in the chapter on chemical treatment.

3.3 Chemical delivery

3.3.1 Weaponisation
The weaponisation of chemical agents involves several steps, for example, the addition of:

◆ **Stabilisers** to prevent the degradation of agents

◆ **Thickeners** to increase the viscosity and persistence of liquid agents

◆ **Carriers** to improve the dispersion of the agent

Following production, the agent must be inserted into the appropriate munitions to achieve desired effects on the target. Chemical munitions are designed to convert their payload into an aerosol of microscopic droplets or particles (which can be readily absorbed by the lungs) or into a spray of larger droplets (which can contaminate large areas or penetrate the skin).

Many munitions and dissemination methods exist. These include explosive, thermal, pneumatic or mechanical means. A spray tank can be used to disseminate agents from aircraft in the same way that a crop duster can disseminate insecticides or pesticides. The same type of ground-based aerosol generator, used to disseminate pesticides, can also be used for chemical agent dispersion.

One means of obtaining a 'pre-packed' chemical weapon, or military-grade chemical material, is by acquiring an item of military ordnance. Despite the long list of signatories to the Chemical Weapons Convention (CWC) such items may still feature in the inventories of armed forces around the world, or possibly be recovered from battlefields or training areas (This is especially the case in relation to countries supplied or assisted by the Former Soviet Union [FSU]).

Military munitions, including artillery shells, aerial bombs (including cluster bombs), spray tanks, missiles, rockets, mines and grenades could each contain chemical materials. All of these munitions types are intended to:

◆ Provide an appropriately sized aerosol (1 to 7 μm) that will remain suspended in the air close to the ground (2 to 3 m), where it will be readily inhaled

◆ Provide a coarse spray that will remain on the ground and cause terrain contamination and degrade human performance due to the necessity of wearing of protective clothing

Most of the standard chemical munitions contain a 'burster' charge surrounded by the agent. The burster charge, when activated by an appropriate fuse, ruptures the munition and causes the agent to be disseminated. Optimum fusing can vary depending on the agent. Impact fusing, employed in ground-burst munitions, is usually used in conjunction with volatile, non-persistent agents since these will generally dissipate if disseminated at too high an altitude. Proximity fusing (barometric pressure or timer activated) is usually used in conjunction with persistent agents which can be disseminated at higher altitudes and still reach the target.

It is increasingly possible that military munitions may be acquired by paramilitary or terrorist groups. The following list is intended to provide first responders with an indication of what may be encountered to assist with (a) recognition, and (b) the passage of information to bomb disposal assets. It is not a definitive list and should not be treated as such.

Artillery rounds:
105 mm projectile (NATO)

---

155 mm projectile (NATO)
8 in projectile (NATO)

These can be filled with nerve agents or mustard. They are bulk filled and dissemination is accomplished by an explosive charge. They weigh between 44 and 197 lb (20 and 89 kg) and contain between 3 and 15 lb (1.4 and 7 kg) of agent, depending on their size.

122 mm (FSU)
130 mm (FSU)
152 mm (FSU)

The 130 mm rounds contained 1.6 kg sarin or 1.4 kg VX. Two 152 mm rounds were produced. One contained 2.8 kg sarin, the other 5.4 kg of blister agent.

130 mm (Iraq)
The Iraqi round was designed to contain a blister agent. The exact quantity is not known, but is likely to have been in excess of 1.5 kg.

Mortar rounds:
4.2 in mortar round (NATO)

This may be filled with mustard agent. It weighs 25 lb (11 kg) and can accommodate approximately 6 lb (2.7 kg) of agent. The FSU and allies of the FSU possessed similar weapons.

Rockets:

The US inventory contained one rocket, the 115-mm M55, which was filled with nerve agent. Each rocket weighs 58 lb (26 kg) and contains about 10 lb (4.5 kg) of agent. The FSU and allies of the FSU possessed similar weapons including 122 mm rockets and 240 mm rockets.

Air dropped weapons:

There were several bulk-filled bombs in the NATO inventory, ranging from the Mk 94 (which contained 108 lb [49 kg] of nerve agent GB) through the MC-1 (which contained 220 lb [99 kg] of agent) to the Mk 116 Weteye (which carried 347.5 lb [156 kg] of nerve agent GB). All of these bombs disseminate the agent by means of an explosive burster charge inside the munition.

The FSU and allies of the FSU possessed similar weapons, including the KhAB-200 and KhAB-500. These weapons could accommodate persistent or non-persistent agents. They were just under 2 m in length (measured without fins) and either 0.3 m or 0.45 m in diameter.

Land mines:

M 23 (NATO), which contained 10.5 lb (4.7 kg) of nerve agent VX. This mine could be used in anti-personnel or anti-vehicle mode.

M1 (NATO), similar in appearance and operation to a one gallon *blast incendiary*. Typically, the M1 contained a blister agent.

The FSU and allies of the FSU possessed similar weapons.

Miscellaneous:
Grenades – Some types of grenade are classified as 'chemical'. Typically, such ordnance will contain riot control agents (for example, CS or CN) or possibly smoke compositions. Hand-thrown grenades are most unlikely to contain lethal materials other than explosives.

Containers – One other item worthy of note is the one-ton container, which was the standard bulk-agent storage and shipping container designed to permit the transfer of agent to other items as needed.

A knowledgeable munitions specialist could recover the agent from any of these items and use it to fill other items more suited to terrorist applications.

3.3.2 Commercial delivery systems
Commercial delivery systems are likely to be less efficient and reliable than munitions and delivery systems specially designed and tested for use with chemical agents. However, such commercial systems can pose a significant threat, especially since many of the technologies are dual-use and dual-purpose.

As noted earlier, a commercial crop-dusting aircraft can serve as a very efficient delivery system for chemical agents, much as a spray tank on a military aircraft would be used to lay down a *line source*. Similarly, individual small generators can disperse chemical agents. A pesticide generator, used to spray crops from a pick-up truck, can be used successfully to contaminate large areas when loaded with an appropriate C or B material. It is known that suitcase generators have been used to contaminate individual rooms or buildings and, on a significantly smaller scale, an *umbrella gun* has been used in at least two assassinations.

Many commercially available dissemination devices are available and could be easily adapted to disperse chemical agents for terrorist purposes. At a more sophisticated level, air- or ground-based aerosol generators can be used for more controlled dissemination of CW agents. These systems may be extremely difficult to detect before use, but measures must be developed to identify *suspicious items* as quickly and as safely as possible.

As noted earlier, simple dissemination systems can be adapted from commercially available systems (most of these devices are dual purpose). It is important to appreciate **the small quantity of chemical agent required to inflict a large number of casualties**, especially to unwarned and unprotected personnel. (For example, many military-developed artillery shells contain less than 10 kg of nerve agent, yet have the potential to injure or kill many hundreds (or possibly thousands) of people.)

It must be noted that **in the very few documented cases of CW terrorism**, such as Aum Shinrikyo in Tokyo, **the quantities of nerve agent used were relatively small and the quality was not high.** Similarly, the dissemination devices were crude and inefficient, depending primarily on evaporation to disseminate the agent into an enclosed space. This combination of factors led to significantly smaller numbers of casualties than may otherwise have been the case. **Simple aerosol generators, such as a pressurised spray can, could serve as relatively effective dissemination systems**. Air-pressure sprayers of the pump-up variety, available at garden centres and DIY stores, could also be utilised. While home-made systems may not be as effective as their military counterparts, the **potential** for mass casualties, if the terrorist gets it right, is enormous.

Of greater potential concern is the use of aerosol generators mounted on the bed of a truck or small watercraft. Commercial generators could disseminate a line-source of agent upwind of a city, a business centre or other populated area. Globally, and on a very small scale, several crude attempts have been made to contaminate individual buildings by the insertion of chemical materials through ventilation systems.

Fortunately, most of these attempts have not been very successful. However, first responders should be able to recognise such makeshift dissemination systems and be prepared to report the salient details to EOD officers.

3.3.3 Meteorological conditions

Meteorological conditions determine the success or failure of a chemical attack.

◆ Temperature of the air and the ground significantly affect agent characteristics. Higher air temperatures may cause the evaporation of aerosol particles, decreasing their size and increasing the likelihood that they will reach the lungs. The temperature of the ground either increases or decreases evaporation rates, thereby increasing or decreasing the duration of contamination. For example, contamination of bare ground by unpurified liquid mustard at a concentration of 30 g/m^2, will persist for several days at temperatures below 10°C (50°F). Conversely, the same amount of agent will last only 1 to 2 days at 26°C (80°F).

◆ Humidity is also known to affect the dissemination (and effectiveness) of a chemical material. High relative humidity may lead to the enlargement of aerosol particles, thereby reducing the quantity of inhalable aerosol. However, the combination of high temperature and high humidity causes increased perspiration in humans, thus intensifying the cutaneous effects of mustard agent and accelerating the transfer of nerve agents through the skin.

◆ Various types of atmospheric precipitation also influence the effects of a chemical attack:
  - Light rain disperses and spreads a chemical agent, thus presenting a larger surface for evaporation and raising the rate of evaporation
  - A heavy rain will dilute and disperse a chemical agent, thereby facilitating its penetration into the ground
  - Heavy rains can also wash away liquid agents, thus causing unintended contamination of non-target areas
  - Snow increases the persistence of contamination by slowing the evaporation rate
◆ Windspeed and direction are critical in determining the success of a chemical attack. Windspeed determines how fast a primary cloud moves and the direction determines the area that must be alerted *downwind*. High winds can disperse vapours, aerosols and liquids rapidly, thereby shrinking the target area and reducing the population's exposure to an agent. High winds also necessitate the use of larger amounts of chemical agent to be effective. Chemical clouds are most effective when winds are steady and less than 4 km/H.

The distance that vapour will travel from contaminated ground is a function of the soil type and the windspeed. Agents may persist in sandy soil three times as long as in a clay soil. **Chemical agents may persist longer in an urban environment than in an open environment since building materials are often porous and**

readily absorb agents, which will be slowly released later.

The nature of buildings and terrain can have a significant impact on a chemical attack in other ways. Greater turbulence of the primary cloud due to woodland and hilly terrain gives shorter distances for cloud travel. In some instances, however, these terrain variations may lengthen the effects, since the chemical will be retained in the area. In ordinary buildings, protection can be improved by closing doors and windows, turning off ventilation systems and sealing all cracks with tape or wet towels.

Persistence:
Unpurified liquid mustard (30 g/m^2 concentration)
< 50°F (10°C)   Several days
80°F (26°C)   1 to 2 days

Factors that influence the effects of a chemical attack:
◆   Temperature (both air and ground)
◆   Humidity
◆   Precipitation
◆   Windspeed
◆   Nature of buildings and terrain

3.4 Nerve agents
Nerve agents - how they work
◆   Nerve agents interfere with transmission of the
    message from nerve to organ

◆ The nerve is normal; the transmission to the organ (muscle, gland) is faulty
◆ The organ (muscle, gland) gets the wrong message, and does the wrong thing. This causes too much activity in muscles, glands.

Effects of nerve agent liquid on the skin
◆ Very small drop: sweating, twitching at site
◆ Small drop: nausea, vomiting, diarrhoea
◆ Drop: loss of consciousness, convulsions, breathing stops, flaccid paralysis

Effects begin within 30 minutes (large amount) to 18 hours (small amount).

Effects of nerve agent vapour
◆ Small amount:
  • Eyes: small pupils, red conjunctiva, dim/blurred vision, pain, nausea/vomiting
  • Nose: runny nose
  • Mouth: increased salivation
  • Airways: tightness in chest, shortness of breath, cough
◆ Large amount:
  • Loss of consciousness
  • Convulsions
  • Flaccid paralysis
  • Breathing stops
  • Heart stops

Effects begin within seconds to a minute.

Tabun (GA): Chemical and physical properties

Boiling point	446°F (230°C)
Vapour pressure	0.037 mm Hg at 68°F (20°C)

Density:

Vapour (compared to air)	5.6
Liquid	1.08 g/ml at 77°F (25°C)
Volatility	610 mg/m³ at 77°F (25°C)
Appearance	Colourless to brown liquid
Odour	Fairly fruity

Solubility:

In water	9.8 g/100 g at 77°F (25°C)
In other solvents	Soluble in most organic solvents

Source: Sidell, F.R., Takafuji, E.T., Franz, D.R. Medical Aspects of Chemical and Biological Warfare, Part I. Washington, DC: Office of the Surgeon General at TMM Publications, Borden Institute, Walter Reed Army Medical Center, 1997.

Tabun (GA): Environmental and biological properties

Detectability:

Vapour	For example, many commercially available detectors (see section 8.1)
Liquid	Detector paper

Persistence:

In soil	Half-life 1-1.5 days
On material	Unknown
Decontamination of skin	M258A1, diluted hypochlorite, soap and water, M291 kit

Biologically effective amount:

Vapour (mg min/m^3)	Ct_{50}: 2-3 (miosis)
	LCt_{50}: 400
Liquid	LD_{50}: (skin): 1.0 g/70 kg man

Source: Sidell, F.R., Takafuji, E.T., Franz, D.R. Medical Aspects of Chemical and Biological Warfare, Part I. Washington, DC: Office of the Surgeon General at TMM Publications, Borden Institute, Walter Reed Army Medical Center, 1997.

Sarin (GB): Chemical and physical properties

Boiling point   316°F (158°C)
Vapour pressure   2.1 mm Hg at 68°F (20°C)

Density:
Vapour (compared to air) 4.86
Liquid   1.10 g/ml at 68°F (20°C)
Volatility   22,000 mg/m³ at 77°F
                                 (25°C)

Appearance   Colourless liquid
Odour   No odour

Solubility:
In water   Miscible
In other solvents   Soluble in all solvents

Source: Sidell, F.R., Takafuji, E.T., Franz, D.R.
Medical Aspects of Chemical and Biological
Warfare, Part I. Washington, DC: Office of the
Surgeon General at TMM Publications, Borden
Institute, Walter Reed Army Medical Center, 1997.

Sarin (GB): Environmental and biological properties

Detectability:

Vapour	For example, many commercially available detectors (see section 8.1)
Liquid	Detector paper

Persistence:

In soil	2-24 hours at 41°F - 77°F (5°C - 25°C)
On material	Unknown
Decontamination of skin	M258A1, diluted hypochlorite, soap and water, M291 kit

Biologically effective amount:

Vapour (mg min/m³)	Ct_{50}: 3 (miosis)
	LCt_{50}: 100
Liquid	LD_{50}: (skin): 1.7 g/70 kg man

Source: Sidell, F.R., Takafuji, E.T., Franz, D.R. Medical Aspects of Chemical and Biological Warfare, Part I. Washington, DC: Office of the Surgeon General at TMM Publications, Borden Institute, Walter Reed Army Medical Center, 1997.

Soman (GD): Chemical and physical properties

Boiling point	388°F (198°C)
Vapour pressure	0.40 mm Hg at 77°F (25°C)

Density:
Vapour (compared to air)	6.3
Liquid	1.02 g/ml at 77°F (25°C)
Volatility	3,900 mg/m³ at 77°F (25°C)
Appearance	Colourless liquid
Odour	Fruity; oil of camphor

Solubility:
In water	2.1 g/100 g at 68°F (20°C)
In other solvents	Soluble in some solvents

Source: Sidell, F.R., Takafuji, E.T., Franz, D.R. Medical Aspects of Chemical and Biological Warfare, Part I. Washington, DC: Office of the Surgeon General at TMM Publications, Borden Institute, Walter Reed Army Medical Center, 1997.

Soman (GD): Environmental and biological properties

Detectability:

Vapour	For example, many commercially available detectors (see section 8.1)
Liquid	Detector paper

Persistence:

In soil	Relatively persistent
On material	Unknown
Decontamination of skin	M258A1, diluted hypochlorite, soap and water, M291 kit

Biologically effective amount:

Vapour (mg min/m³)	Ct_{50}: 2-3 (miosis) LCt_{50}: 50
Liquid	LD_{50}: (skin): 350 mg/70 kg man

Source: Sidell, F.R., Takafuji, E.T., Franz, D.R. Medical Aspects of Chemical and Biological Warfare, Part I. Washington, DC: Office of the Surgeon General at TMM Publications, Borden Institute, Walter Reed Army Medical Center, 1997.

VX: Chemical and physical properties

Boiling point	568°F (298°C)
Vapour pressure	0.0007 mm Hg at 68°F (20°C)

Density:
Vapour (compared to air)	9.2
Liquid	1.008 g/ml at 68°F (20°C)
Volatility	10.5 mg/m^3 at 77°F (25°C)
Appearance	Colourless to straw-coloured liquid
Odour	Odourless

Solubility:
In water	Miscible <48.9°F (9.4°C)
In other solvents	Soluble in all solvents

Source: Sidell, F.R., Takafuji, E.T., Franz, D.R. Medical Aspects of Chemical and Biological Warfare, Part I. Washington, DC: Office of the Surgeon General at TMM Publications, Borden Institute, Walter Reed Army Medical Center, 1997.

VX: Environmental and biological properties

Detectability:

Vapour	For example, many commercially available detectors (see section 8.1)
Liquid	detector paper

Persistence:

In soil	2-6 days
On material	Persistent
Decontamination of skin	M258A1, diluted hypochlorite, soap and water, M291 kit

Biologically effective amount:

Vapour (mg min/m^3)	Ct_{50}: 10-50 (death)
	LCt_{50}: 10
Liquid	LD_{50}: (skin): 10 mg/70 kg man

Source: Sidell, F.R., Takafuji, E.T., Franz, D.R. Medical Aspects of Chemical and Biological Warfare, Part I. Washington, DC: Office of the Surgeon General at TMM Publications, Borden Institute, Walter Reed Army Medical Center, 1997.

3.4.1 Early recognition

The Tokyo example – The effects of nerve agent vapour appear almost immediately. In a large group of exposed people, the effects will vary in intensity. In the Tokyo subway incident, about 75 per cent of the casualties who were admitted to medical facilities had minor effects and probably did not require any therapy. (Indeed, some research suggests that many people who self-presented could be classed as the worried well – a phenomenon referred to be some as 'sarinoia'.) Conversely, a small number of those affected died before receiving medical treatment.

General indicators – After nerve agent vapour exposure, almost all people affected will have small pupils (miosis), usually in both eyes but occasionally unilaterally. **Since even unaffected persons have small pupils in bright sunlight, possible nerve agent victims must be examined in a darkened area such as a shady wooded area or in a darkened vehicle**. Many will also have a runny nose (rhinorrhoea) and many will complain about some degree of shortness of breath. **These three conditions - small pupils, runny nose and shortness of breath - may be mild or they might be very severe when the casualty is first seen. Some casualties may have only one of these effects, some may have two and some may have all three**.

A large concentration of vapour will cause sudden loss of consciousness and convulsions; a small amount will also cause these effects if the casualty remains in the vapour for many minutes.

◆ After a short period of convulsions, the casualty will stop breathing and become paralysed and limp

◆ These severe casualties will have small pupils, copious secretions from the nose and mouth. (In one reported incident, a casualty stated that his mask 'filled' as a result of nasal secretion).

◆ Unless they are entirely limp, will have **fasciculations (which look like ripples or worms moving under the skin)**

These three signs distinguish between a casualty convulsing from nerve agent poisoning and one convulsing from cyanide poisoning. The casualty may also have cyanosis - a bluish colour in the lips, nail beds and possibly skin. This does not occur as readily after cyanide intoxication, though this is not a distinguishing point.

Nerve agents cause biological effects over a wide range of vapour concentrations. Thus, after release of a nerve agent – accidental or otherwise – the casualties will have a range of symptoms: some may have small pupils with no other effects; others may have convulsions, paralysis and cessation of breathing.

3.4.2 Physical examination by medical personnel

A person exposed to a small amount of nerve agent vapour may have **miosis** - which will not be noticed in bright light - **secretions from the nose and mouth**, and **crackles and rales upon auscultation** of the lungs. Not all of these signs may be present in each casualty.

Exposure to a small amount of liquid on the skin may produce no specific signs. There may be **localised fasciculations** and **sweating** if the site of contact is known and examined. **Hyperactive bowel sounds** may accompany gastrointestinal complaints. Miosis is usually not present after liquid exposure on the skin unless the exposure was large or an agent droplet encounters the eye or the skin near the eye.

A severe casualty from either vapour or liquid exposure may be **unconscious and convulsing**, with **miosis**, **copious secretions**, **respiratory distress or apnoea**, and **generalised muscular fasciculations**.

3.4.3 Nerve agent characteristics and properties

Nerve agents are chemicals that disrupt the mechanism by which nerves communicate with the organs they stimulate. This interference with normal communication causes overstimulation of these organs. Many substances produce similar symptoms to those associated with nerve agent exposure. However, these are not typically considered 'nerve agents' by military standards; they are significantly less toxic.

Military nerve agents – Military nerve agents produced in quantity include: tabun (GA), sarin (GB), soman (GD), GF and VX.

When fresh, nerve agents are clear, colourless liquids. Two agents, tabun and soman, are said to have slight nondescript odours. Liquid agents are heavier than water, and their vapour is heavier than air, which means that they will sink into low terrain and basements.

Nerve agents act by disrupting the normal transmission of messages between nerves and their receiving organs. They do this by blocking the activity of an enzyme, acetylcholinesterase, that normally destroys and stops the activity of the chemical messenger or neurotransmitter, acetylcholine. This neurotransmitter is released by a nerve to stimulate a muscle or gland and, when it is not destroyed, it continues to stimulate the muscles or glands, causing hyperactivity. This overstimulation causes the signs and symptoms of nerve agent poisoning.

Major effects from nerve agents occur in (a) the skeletal muscles (those muscles in arms, legs and other parts of the body that can be voluntarily moved) and (b) the smooth muscles (those within the body such as in the airways and in the gastrointestinal tract). The glands that can be affected include those that secrete to the outside of the body - that is, in the nose (rhinorrhoea), in the mouth (salivation), in the eyes (tearing) and in the skin (sweating) and those that secrete inside the body - for example, in the walls of the airways and gastrointestinal tract.

Exposure to nerve agent vapour initially affects the sensitive organs of the face that come into direct contact with the vapour: the eyes, nose, mouth and airways. The pupils become small (miosis), the eyes become red and the casualty may complain of dim vision (primarily because small pupils do not allow light into the eye), blurred vision, pain in the eye or head, and nausea and vomiting. Exposure of the nose causes secretions (rhinorrhoea). Exposure of the mouth causes excessive salivation. If the agent is inhaled, the airways become smaller or constricted and the casualty will have shortness of breath and coughing, similar to an asthma sufferer. These effects begin within seconds of contact with vapour.

A few seconds after inhalation of a large amount of nerve agent vapour, there will be a sudden loss of consciousness followed by convulsions. If no care is given within a few minutes the casualty will stop breathing and become flaccid or completely limp.

Effects from liquid may begin within 30 minutes (in the case of a large amount) **and any time up to 18 hours later** (in the case of a small amount). A very small amount of nerve agent liquid on the skin will first produce sweating and a small amount of muscular twitching (fasciculations) at the site of the drop. A slightly larger drop will cause nausea and vomiting. These effects from small droplets may not begin until hours after contact with the agent. A larger, lethal-sized drop will, within a few minutes of contact, cause loss of consciousness, convulsions, cessation of respiration and paralysis.

3.5 Cyanide
◆ Cyanide poisons cells (stops use of oxygen)
◆ The cell cannot use oxygen and it dies
◆ Oxygen remains in the blood (blood stays red)

Effects of cyanide
◆ Small amount: no effects
◆ Medium amount: dizziness, nausea, feeling of weakness
◆ Large amount:
◆ Loss of consciousness
◆ Convulsions
◆ Breathing stops
◆ Death
◆ First effect: seconds

Effects of Cyanogen Chloride
◆ Small amount: irritation; giddiness, nausea, feeling of weakness
◆ Large amount: unconsciousness, convulsions

Hydrogen Cyanide (AC): Chemical and physical properties

Boiling point	78°F (25.7°C)
Vapour pressure	740 mm Hg
Density:	
Vapour	0.99 at 68°F (20°C)
Liquid	0.68 g/ml at 77°F (25°C)
Solid	NA
Volatility	1.1×10^6 mg/m³ at 77°F (25°C)
Appearance and odour	Gas: odour of bitter almonds or peach kernels
Solubility:	
In water	Complete at 77°F (25°C)
In other solvents	Completely miscible in almost all organic solvents

Source: Sidell, F.R., Takafuji, E.T., Franz, D.R. Medical Aspects of Chemical and Biological Warfare, Part I. Washington, DC: Office of the Surgeon General at TMM Publications, Borden Institute, Walter Reed Army Medical Center, 1997.

Hydrogen Cyanide (AC): Environmental and biological properties

Detection

For example, many commercially available detectors (see section 8.1)
(detector paper for liquid)

Persistence:
In soil   <1 hour
On material   Low
Skin decontamination   Water; soap and water

Biologically effective amount:
Vapour (mg min/m^3) Ct_{50}: 2,500-5,000 (death)
                                LCt_{50}: 2,500-5,000 (time -dependent)
Liquid LD_{50} (skin): 100 mg/kg

Source: Sidell, F.R., Takafuji, E.T., Franz, D.R. Medical Aspects of Chemical and Biological Warfare, Part I. Washington, DC: Office of the Surgeon General at TMM Publications, Borden Institute, Walter Reed Army Medical Center, 1997.

Cyanogen Chloride (CK): Chemical and physical properties

Boiling point	55°F (12.9°C)
Vapour pressure	1,000 mg Hg
Density:	<1 hour
Vapour	2.1
Liquid	1.18 g/ml at 68°F (20°C)
Solid	Crystal: 0.93 g/ml at −104°F (−40°C)
Volatility	2.6×10^6 mg/m³ at 55°F (12.9°C)
Appearance and odour	Colourless gas or liquid
Solubility:	
In water	6.9 g/100 ml at 68°F (20°C)
In other solvents	Most organic solvents (mixtures are unstable)

Source: Sidell, F.R., Takafuji, E.T., Franz, D.R. Medical Aspects of Chemical and Biological Warfare, Part I. Washington, DC: Office of the Surgeon General at TMM Publications, Borden Institute, Walter Reed Army Medical Center, 1997.

Cyanogen Chloride (CK): Environmental and biological properties

Detection

Persistence:
In soil	Non-persistent
On materiel	Non-persistent
Skin decontamination	Water; soap and water

Biologically effective amount:
Vapour (mg min/m³)	Ct_{50}: 11,000 (death)
	LCt_{50}: 11,000
Liquid	NA

Source: Sidell, F.R., Takafuji, E.T., Franz, D.R. Medical Aspects of Chemical and Biological Warfare, Part I. Washington, DC: Office of the Surgeon General at TMM Publications, Borden Institute, Walter Reed Army Medical Center, 1997.

3.5.1 Early recognition

In contrast to nerve agents, **cyanide**, although commonly considered a very deadly chemical, **causes almost no effects after brief exposure to low concentrations**. There are no distinguishing physical signs after a small exposure, and symptoms regress after the casualty has been in clean air for some time. **Cyanide in moderate amounts may produce nausea and feelings of dizziness, weakness and anxiety. A large concentration of cyanide will produce serious effects, including loss of consciousness within seconds.** This is followed by several minutes of **convulsions** before the **casualty stops breathing**. Loss of consciousness and death may occur even after a very brief exposure to a large concentration. **As with nerve agents, if a person remains exposed to a low concentration for many minutes, he/she will also have severe effects**.

Another form of cyanide, **cyanogen chloride, will cause almost immediate irritation to the eyes, nose and airways**, and these effects may be difficult to distinguish from those caused by a riot control agent. If the exposure to cyanogen chloride is large, the casualty will convulse within seconds, as happens after exposure to cyanide. If the exposure to cyanogen chloride is small this irritation may be the only effect and no further action is necessary.

3.5.2 Physical examination by medical personnel

People convulsing or who have convulsed from cyanide usually have normal-sized to large pupils, usually do not have excessive secretions and do not have muscular fasciculations (ripples under the skin) – all of which are seen in nerve agent casualties. On the other hand, **those poisoned with cyanide often have skin that is redder than normal because of the venous blood changing from blue to red. The odour of bitter almonds may be present.**

An individual who is exposed to a smaller amount of cyanide vapour or someone who eats or drinks cyanide (to commit suicide) will experience the progression of events more slowly. The casualty may feel giddy or dizzy, may be nauseated and may feel weak. It may be many minutes before loss of consciousness. There are no specific physical signs.

3.5.3 Cyanide properties

Cyanide is a common chemical. Not only does it occur naturally in some foods (the pits of some fruits and lima beans, for example), but it is also manufactured commercially as cyanide salts: sodium cyanide, potassium cyanide and calcium cyanide. Cyanide is used in ore extraction, in tanning, in electroplating and in many chemical syntheses, including the manufacture of paper, textiles and plastics. In addition, cyanide is in the smoke of burning plastics and other synthetic materials. People who work in these industries are at risk from cyanide exposure, and firefighters may be exposed to it in the course of fighting fires involving such materials.

The military has used two forms of cyanide - hydrocyanic acid (hydrogen cyanide) and cyanogen chloride. The cyanides are erroneously called 'blood agents' by the military; they do not cause effects or have other activity in blood. They are carried in the blood from their point of absorption to other organs, though other agents are similarly transported and are not referred to as blood agents.

Hydrocyanic acid and cyanogen chloride are very volatile liquids - that is, they evaporate quickly and become vapours or gases. Hydrocyanic acid vapour is lighter than air and will rise from the ground. The vapour of cyanogen chloride is heavier than air and will sink into low terrain and basements. Hydrocyanic acid has the odour of bitter almonds; because of genetic factors only about half the population can smell it.

Its volatility, the large amount required (dose) - at least compared to other CW agents - and the fact that it is lighter than air, make cyanide a less-than-ideal agent outdoors. However, it would be a good agent in confined spaces.

Hydrocyanic acid can be very easily made by mixing a cyanide salt with a strong acid; these chemicals, and a device to mix them, were found in Tokyo subway restrooms a few weeks after the nerve agent release.

Cyanogen chloride is a compound that is converted to cyanide as soon as it enters the body. It causes the same effects as cyanide, but in addition it causes irritation of the eyes, nose and mucous membranes, much as riot control agents.

Cyanide produces clinical effects by causing cell death. It does this by entering each cell of the body and poisoning the mechanism that uses oxygen. Oxygen enters the body through the lungs and is carried by the blood to the cells. Cyanide prevents the cells from using the oxygen and they suffocate. As a result, oxygen remains in the blood and the bright red blood that is carried in arteries does not change; the blood in veins, which normally shows through the skin as bluish, appears bright red also. Since cells cannot use oxygen, they turn to metabolism without oxygen, or anaerobic metabolism, which results in acidosis if the person lives long enough.

The human body can destroy, or detoxify, a small amount of cyanide, although this amount is limited because the body does not have enough sulphur to react with cyanide. Because of this, cyanide does not cause effects when a person is exposed to a small dosage. This is in contrast to some other CW agents, which cause effects at very low amounts.

The brain, or central nervous system, is very dependent on oxygen, and most effects of cyanide poisoning are those caused by a lack of oxygen in

the brain. After exposure to a large amount of cyanide vapour, there is a sudden loss of consciousness, followed within seconds by convulsions. In about three to five minutes convulsions stop because breathing stops. The casualty is unconscious, paralysed and not breathing for several minutes before the heart stops. Death will occur within 10 minutes.

3.6 Blister agents (vesicants)

Mustard - how it works
◆ Mustard quickly penetrates the skin, mucous membranes (eye, airways)
◆ It changes to another substance, and reacts with enzymes, proteins, DNA
◆ It causes cell death
◆ Mustard effects are like radiation ('radiomimetic')
◆ Mustard causes damage within minutes
◆ No mustard is in blister fluid

Early effects of blister agents
◆ Mustard: no effects for hours then reddening and blistering of the skin
◆ Lewisite, phosgene oxime: irritation, pain; these do not improve with fresh air and get worse with time

Effects of Lewisite and Phosgene Oxime
◆ Very irritating, pain on contact
◆ Tissue damage evident within minutes
◆ Damage to eyes, skin, airways similar to that caused by mustard

---

Impure Sulphur Mustard (H): Chemical and physical properties

Boiling point	Varies
Vapour pressure	Depends on purity
Density:	
Vapour (compared to air)	Approx 5.5
Liquid	Approx 1.24 g/ml at 77°F (25°C)
Solid	N/A
Volatility	Approx 920 mg/m^3 at 77°F (25°C)
Appearance	Pale yellow to dark brown liquid
Odour	Garlic or mustard
Solubility:	
In water	0.092 g/100 g at 72°F (22°C)
In other solvents	Complete in CCl$_4$, acetone, other organic solvents

Source: Sidell, F.R., Takafuji, E.T., Franz, D.R. Medical Aspects of Chemical and Biological Warfare, Part I. Washington, DC: Office of the Surgeon General at TMM Publications, Borden Institute, Walter Reed Army Medical Center, 1997.

Impure Sulphur Mustard (H): Environmental and biological properties

Detection   Liquid: detector paper
                                   Vapour: Many
                                   commercially available
                                   detectors (see section
                                   8.1)

Persistence:
In soil   Persistent
On materiel   Temperature-dependent;
                                   hours to days
Decontamination of skin   M258A1, dilute
                                   hypochlorite, soap and
                                   water, M291 kit

Biologically effective amount:
Vapour (mg min/m^3) LCt_{50}: 1,500
Liquid LD_{50}: approx 100 mg/kg

Source: Sidell, F.R., Takafuji, E.T., Franz, D.R. Medical Aspects of Chemical and Biological Warfare, Part I. Washington, DC: Office of the Surgeon General at TMM Publications, Borden Institute, Walter Reed Army Medical Center, 1997.

Distilled Sulphur Mustard (HD): Chemical and physical properties

Boiling point	441°F (227°C)
Vapour pressure	0.072 mm Hg at 68°F (20°C)
Density:	
Vapour (compared to air)	5.4
Liquid	1.27 g/ml at 68°F (20°C)
Solid	Crystal: 1.37 g/ml at 68°F (20°C)
Volatility	610 mg/m³ at 68°F (20°C)
Appearance	Pale yellow to dark brown liquid
Odour	Garlic or mustard
Solubility:	
In water	0.092 g/100 g at 72°F (22°C)
In other solvents	Complete in CCl₄, acetone, other organic solvents

Source: Sidell, F.R., Takafuji, E.T., Franz, D.R. Medical Aspects of Chemical and Biological Warfare, Part I. Washington, DC: Office of the Surgeon General at TMM Publications, Borden Institute, Walter Reed Army Medical Center, 1997.

Distilled Sulphur Mustard (HD): Environmental and biological properties

Detection	Liquid: detector paper Vapour: Many commercially available detectors (see section 8.1)
Persistence: In soil On materiel	 2 weeks - 3 years Temperature-dependent; hours to days
Decontamination of skin	M258A1 kit, dilute hypochlorite, soap and water, M291 kit

Biologically effective amount:

Vapour (mg min/m^3)	Ct_{50}: 12-200 (eye) Ct_{50}: 100-200 (pulmonary) Ct_{50}: 200-1,000 (erythema) LCt_{50}: 1,500 (inhalation) LCt_{50}: 10,000 (skin)
Liquid on skin	10 μg (erythema) LD_{50}: 100 mg/kg

Source: Sidell, F.R., Takafuji, E.T., Franz, D.R. Medical Aspects of Chemical and Biological Warfare, Part I. Washington, DC: Office of the Surgeon General at TMM Publications, Borden Institute, Walter Reed Army Medical Center, 1997.

CHAPTER 3: Chemical Agents

Lewisite (L): Chemical and physical properties

Boiling point	374°F (190°C)
Vapour pressure	0.39 mm Hg at 68°F (20°C)

Density:
Vapour (compared to air)	7.1
Liquid	1.89 g/ml at 68°F (20°C)
Solid	NA
Volatility	4,480 mg/m^3 at 68°F (20°C)
Appearance	Pure: Colourless, oily liquid
	As agent: amber to dark brown liquid
Odour	Geranium

Solubility:
In water	Slight
In other solvents	Soluble in all common organic solvents

Source: Sidell, F.R., Takafuji, E.T., Franz, D.R. Medical Aspects of Chemical and Biological Warfare, Part I. Washington, DC: Office of the Surgeon General at TMM Publications, Borden Institute, Walter Reed Army Medical Center, 1997.

Lewisite (L): Environmental and biological properties

Detection   Vapour: Many
                                   commercially available
                                   detectors (see section 8.1)
                                   Liquid: detector paper

Persistence:
In soil   Days
On materiel   Temperature-dependent;
                                   hours to days
Skin decontamination   Dilute hypochlorite,
                                   M258A1 kit, water, M291
                                   kit

Biologically effective amount:
Vapour (mg min/m³)   Eye: < 30
                                   Skin: approx 200
                                   Ct_{50}: 1,500 (erythema)
                                   LCt_{50}:   1,200-1,500
                                   (inhalation)
Liquid on skin   10-15 µg
                                   40-50 mg/kg

Source: Sidell, F.R., Takafuji, E.T., Franz, D.R. Medical Aspects of Chemical and Biological Warfare, Part I. Washington, DC: Office of the Surgeon General at TMM Publications, Borden Institute, Walter Reed Army Medical Center, 1997.

Phosgene Oxime (CX): Chemical and physical properties

Boiling point	262°F (128°C)
Vapour pressure	11.2 mm Hg at 77°F (25°C) (solid)
	13 mm Hg at 104°F (40°C) (liquid)
Density:	
Vapour (compared to air)	<3.9 (estimated)
Liquid	N/A
Solid	N/A
Volatility	1,800 mg/m³ at 68°F (20°C)
Appearance	Colourless, crystalline solid or a liquid
Odour	Intense, irritating
Solubility:	
In water	70 per cent
In other solvents	Very soluble in most organic solvents

Source: Sidell, F.R., Takafuji, E.T., Franz, D.R. Medical Aspects of Chemical and Biological Warfare, Part I. Washington, DC: Office of the Surgeon General at TMM Publications, Borden Institute, Walter Reed Army Medical Center, 1997.

Phosgene Oxime (CX): Environmental and biological properties

Detection   Many commercially
                                   available detectors (see
                                   section 8.1)

Persistence:
In soil   2 hours
On materiel   Non-persistent
Skin decontamination   Water

Biologically effective amount:
Vapour* Ct_{50}: 200 (eye)
(mg min/m³) Ct_{50}: 2,500 (erythema)
                                   LCt_{50}: 3,200
Liquid on skin   No estimate

Vapour levels are estimated.
Source: Sidell, F.R., Takafuji, E.T., Franz, D.R.
Medical Aspects of Chemical and Biological Warfare,
Part I. Washington, DC: Office of the Surgeon
General at TMM Publications, Borden Institute,
Walter Reed Army Medical Center, 1997.

3.6.1 Early recognition

Mustard

Sulphur mustard produces no effects for hours after exposure, although those exposed might notice the odour of mustard, onion or garlic. **Many hours after exposure** - from 2 to 24 hours - the **exposed person will notice eye irritation, perhaps burning on the skin, or effects of upper airway irritation**. This casualty is not one that will be readily symptomatic for the first responder to recognise. However, first responders should note and set aside for further observation any people who might have been exposed to liquid or vapour contamination.

Lewisite and Phosgene Oxime

The two blister agents that produce immediate effects, Lewisite and phosgene oxime, are uncommon chemicals. **The vapour of these substances causes** effects like those of riot control agents - **burning or pain in the eyes and exposed mucous membranes and on the skin - but the pain from these agents is more severe**. Unlike riot control agents, **this pain does not decrease after the casualty has been in fresh air for a few minutes, but rather it increases. Also, signs of tissue destruction and death (a greyish look to the skin) may be seen within minutes after exposure**.

3.6.2 Physical examination

Mustard
There are no immediate physical signs of mustard exposure. Some hours after exposure, commonly 4 to 8 hours, redness of the skin (erythema) may appear, there may be reddening in the conjunctiva (mild conjunctivitis) and the patient may have upper respiratory complaints, such as aching sinuses and sore throat. These may gradually progress depending on the amount of exposure.

Lewisite and Phosgene Oxime
Evidence of tissue destruction (greyish epithelium) will be present within minutes after contact with these agents. Conjunctivitis will also be present early.

3.6.3 Blister agent properties
There are three agents categorised as blister agents: mustard, Lewisite and phosgene oxime. Mustard produces no immediate effects, while Lewisite and phosgene oxime produce immediate pain on whatever part of the body is exposed to their liquid or vapour, such as the eyes or skin. This immediate pain alerts the person that he/she is being exposed and he/she should leave the area and decontaminate if possible. There is no such alert from mustard, and the person, having no warning, might remain in the area or might not decontaminate and will become severely exposed. Mustard is absorbed rapidly through skin and

mucous membranes, and chemical effects in cells occur within minutes.

Some effects of the sulphur mustard, or mustard, are widely known from photographs of Iranian casualties in the Iran-Iraq war. These effects were skin burns and skin blisters. However, blister agents cause other effects that are as serious or more serious than those on the skin. They also damage the eyes, the airways and some internal organs. Mustard was the largest producer of chemical casualties in the First World War, but generally was not lethal. For example, fewer than 3 per cent of mustard casualties died and many soldiers returned to duty after about two weeks.

Mustard

Mustard is an oily liquid that is heavier than water; its vapour is heavier than air. It has the odour of mustard, onions or garlic, but the odour can be detected only at concentrations that are close to toxic levels.

Mustard can be absorbed into the body through the eyes, the skin and the airways. Mustard absorption begins within seconds of contact with skin or mucous membranes. As it penetrates, some mustard remains on the absorptive surface, damaging the eyes, the skin, and the lining of the airways. Inside the skin, mustard damages the cells that separate the epidermis (upper layer) from the dermis (lower layer). After this damage occurs, the two layers separate, and the space between

CHAPTER 3: Chemical Agents

becomes a blister. Mustard produces similar effects in the airways and eyes, although blisters like those seen on the skin do not appear. Decontamination of liquid from the skin and eyes must be performed immediately if damage is to be prevented or lessened.

After it penetrates, mustard is quickly changed to another chemical, and this chemical quickly reacts with enzymes and proteins within the body, in ways which are not well understood, to cause damage in the cells it contacts. Mustard is a radiomimetic (like x-ray or radiation) compound, and cellular death is like that caused by radiation. The final stage is damage to DNA, a vital component of cells. Because of all of these changes and chemical reactions, there is no free mustard in blister fluid.

Although mustard is absorbed and causes chemical damage in cells within minutes, the noticeable effects do not appear until later. This time ranges from 2 to 24 hours and is usually about 4 to 8 hours. This time is shorter with large amounts and is shorter after liquid exposure than after vapour exposure. Warm, moist areas of the body are more susceptible than other areas. Once damage has begun, the clinical effects may continue to progress for several days. Bone marrow damage, for example, is not evident for 3 to 5 days after exposure.

The first effect on the skin is redness (erythema) - similar to the redness caused by sunburn - with burning and itching. Over a period of hours, small blisters (vesicles) appear and gradually they combine to form large blisters.

Irritation and redness are usually the first effects noticed in the eye. If the amount of exposure is large, the eyelids may become swollen and inflamed, causing them to shut. Damage to the eye itself may occur with irregularities of the cornea, but this is at a later stage. Casualties complain of not being able to see, but this is usually because the eyes are swollen shut, not because of damage to the eye itself.

Damage to the airways begins in the upper airways with sinus pain, irritation of the nose (perhaps with bleeding), a sore throat, and a hacking cough. If more than a minimal amount is inhaled, there may be voice changes, with hoarseness or loss of voice. If a large amount is inhaled, there is damage to the lower airways, with shortness of breath and a severe productive cough. The shorter the onset time of these lower airway effects, the more ominous the prognosis; survival is unlikely if they begin earlier than 4 hours after exposure (because of the increased likelihood of infection).

Absorption of a large amount of mustard will damage the bone marrow causing a decrease in the number of white blood cells (and thus the ability to fight infection), red blood cells and platelets several days after exposure. Death may occur from overwhelming infection despite antibiotic use. Days after absorption of a very large amount of mustard, there may be damage to the gastrointestinal tract with subsequent loss of fluids and electrolytes.

Lewisite and Phosgene Oxime

Lewisite and phosgene oxime, in both their liquid and vapour forms, cause moderate to severe pain on contact with skin or the mucous membranes of the eyes, nose, mouth and airways. They also produce visible greyish tissue damage within several minutes of contact. Later, severe damage of the skin, eyes and airways may appear. Neither Lewisite nor phosgene oxime damages bone marrow. Lewisite, however, causes leakage of systemic capillaries, and hypovolaemia and hypotension may result.

3.7 Pulmonary agents

Effects of pulmonary agents

◆ Shortness of breath (dyspnoea) - at first with exertion, later at rest
◆ Cough - initially hacking cough, later with frothy sputum

Effects begin 2 to 24 hours after exposure.

3.7.1 Early recognition

After exposure to a high concentration of some pulmonary agents (for example, phosgene, chlorine), there may be irritation of the eyes, nose and airways. This, like the irritation caused by riot control agents, will decrease once the casualty is in clean air. Distinguishing a casualty of a pulmonary agent from a person exposed to a riot control agent may be difficult at this point. If a pulmonary agent is suspected, this casualty should be closely watched under medical supervision and kept at absolute rest (no walking) because hours later he may become short of breath from fluid in the lungs.

3.7.2 Physical examination

The casualty will appear short of breath and may be coughing up clear frothy fluid. On auscultation there will be crackles and rales, initially at the bases and later throughout the lung fields.

3.7.3 Pulmonary agent properties

A pulmonary or lung agent, is a chemical that damages the membranes in the lung that separate the alveolus (air sac) from the capillary. As a result of this damaged membrane, the plasma from the blood leaks into the alveoli, filling them with fluid and preventing air from entering. A person with this type of poisoning does not get enough oxygen and dies from suffocation - very similar to drowning. For this reason, this type of poisoning is sometimes called 'dry land drowning' and is also known as non-cardiac pulmonary oedema.

Pulmonary oedema takes hours to develop and the time of onset of the first clinical effect varies from 2 hours to as long as 24 hours after exposure to the agent. Generally, when the onset time is long, the poisoning is less severe. The onset of symptoms within several hours after contact with the agent suggests that the effects will be severe.

The first symptom is usually shortness of breath, which initially is present only during exertion (even slow walking) and later progresses until it is quite severe even at rest. Coughing, with later production of large amounts of clear, frothy sputum (derived from the blood leaking into the alveoli), usually accompanies the shortness of breath.

Fluid loss from the vascular compartment into the lung may be quite large. Hypovolaemia and hypotension are common complications of this loss.

There are a number of common chemicals that cause this type of damage. Phosgene (carbonyl chloride) is the prototype and has been the one most studied. Phosgene was used in the First World War, but it is not considered a chemical warfare agent now. Phosgene evaporates very quickly, and liquid phosgene is not hazardous except as a source of vapour.

Unlike the nerve agents and the vesicants, phosgene causes damage at only one site in the body, namely the alveolar-capillary membrane. To produce this damage, it must be inhaled. Phosgene does not damage the eyes or skin, and a protective mask provides complete protection. Phosgene has the odour of freshly cut grass or newly mown hay, but the concentration for recognition of this odour is close to the concentration that will produce tissue damage.

3.8 Riot control agents

Riot control agents feature in the arsenal of most military and paramilitary organisations. Some variants are also available commercially as self-defence sprays and projectors. While this material is generally considered non-lethal (and quickly dissipates once exposed to air), it is of concern to first responders for two reasons. Firstly, it could panic a civilian population – especially if used in a confined space. Secondly, its effects could be confused with a CW agent.

Effects of riot control agents

◆ Burning, irritation of:
  - Nose, with nasal secretions (rhinorrhoea)
  - Eyes, with reddening of eyes, tearing
  - Mouth, with salivation
  - Airways, with coughing, possibly feeling short of breath
  - Skin, with possible redness (erythema)

Effects start seconds after exposure.

3.8.1 Early recognition

Riot control agents are commonly used in the civilian environment and many people (particularly first responders) are likely to recognise their effects. Within seconds of contact, casualties will complain of burning and stinging in the eyes and nose, and maybe the mouth and skin. They may complain of shortness of breath. Their eyes may be red and teary, and their noses may be running. After evacuation to fresh air, they will begin to feel better. Usually no further first aid is necessary. Any powder on clothing should be removed, and this should be done with care or the rescuer will be exposed and the casualty re-exposed. The unprotected responder who enters an atmosphere containing a riot control agent will begin to feel the effects in a very short time.

3.8.2 Physical examination

A patient exposed to a riot control agent will usually have red eyes, tearing, a runny nose and some degree of respiratory distress. The effects of these agents start within seconds of contact, and **the effects decrease when the casualty moves to clean air**.

3.8.3 Riot control agent properties

The available riot control agents are CS, CN, capsaicin and CR. CS is used by the military and by law enforcement agents. CN is better known as Mace® and is available in small containers for self-protection. Capsaicin is known as "pepper-spray" because it comes from peppers. CR is a British agent. In contrast to the other agents, these agents are solids and are usually dispersed in a liquid spray.

There are minor differences between riot control agents - the effects of capsaicin, for example, last longer than those of CS and CN - but they all produce about the same effects, causing pain or burning on exposed mucous membranes and skin.

Burning in the eyes is usually accompanied by tearing, redness and closure of the eyes. If an individual inhales these substances, there is discomfort in the airways and a feeling of difficulty breathing or of a tight chest. They may also irritate or burn the skin, particularly if the temperature is warm and the skin is moist.

The effects of these agents start within seconds of contact, and the effects decrease when the casualty moves to clean air.

Rarely do these agents produce serious effects. However, a forceful means of dissemination may cause a particle to impact in the eye. Also, they might cause bronchospasm in an individual with hyperactive airways, and a large concentration in high temperature and humidity can cause a delayed (4 to 6 hours until onset) dermatitis with erythema and blisters.

CHAPTER 4: CHEMICAL TREATMENT

CHAPTER 4: CHEMICAL TREATMENT

This chapter is primarily designed to provide advice to trained medical personnel. However, it also provides useful background information for all first responders.

4.1 Nerve agents

Nerve agents of military origin, in terms meaningful to first responders, should be considered odourless, colourless, tasteless and, as such, produce no immediate sensation when in contact with exposed skin. (It is, however, worth noting that the sarin produced by ASK (Aum Shinrikyo) is reported to have had an unpleasant odour.) To minimise the risks associated with a potential nerve-agent incident, first responders should consider the following:

◆ **Protection.** Protect yourself by wearing the appropriate level of Personal Protective Equipment (PPE). This should include both respiratory and dermal protection. Typically, the minimum level of protection for a first responder will include: air purifying respirator, military-type NBC suit or similar Tyvek® garment, suitable gloves, and boots.

◆ **Casualties.** Remove casualty from contamination and contamination from casualty

◆ **Evacuation.** Get the casualty away from the source, such as by moving upwind or out of a contaminated building

◆ **Decontamination.** If it is absolutely certain that exposure was to vapour only, remove outer clothing only

◆ If there is a possibility of liquid contamination, all clothing must be removed and the casualty showered (or washed) with soap and water. Dilute hypochlorite is acceptable in extremes.

◆ **Immediate first aid.** Follow **ABC**s: Management of Airways, Breathing and Circulation. (See below; in relation to a severe casualty, this should follow administration of antidotes).

4.1.1 Triage
Immediate – a casualty requires antidote treatment immediately if they are:
◆ Convulsing
◆ Post-ictal
◆ Not breathing
◆ Has a combination of two or more organ systems affected (breathing difficulty, gastrointestinal effects, skeletal muscular twitching or weakness, disturbances in level of consciousness)

Included in this category (if resources are available to provide appropriate immediate care) would be casualties with no respiratory or cardiac activity.

Delayed – (1) an individual who suspects he or she has or had liquid agent on their skin but who has no other discernible effects, or (2) an individual who was given 4

mg or more of atropine, and who is recovering from the effects of both the agent and the antidote. *Note:* The first type of casualty must be kept under observation for at least 18 hours.

Minimal – a casualty who is able to walk and talk after vapour exposure. Some of these casualties will need treatment (antidotes), others will not. Some will have miosis without other effects and will require no antidotes, others will be short of breath and will require antidotes. Some may be vomiting and therefore antidotes will be required. **No one who is able to walk and talk is usually in danger of immediate loss of life**.

A casualty who is walking and talking after possible liquid exposure may be symptomatic and need antidotes, or may be asymptomatic. In either case, this casualty must be observed for at least 18 hours.

A casualty from a nerve agent who still has cardiac activity (even in the absence of spontaneous respiration) should be triaged as an immediate priority. In one instance, a casualty who had no cardiac activity when first seen by medical personnel survived to walk out of the hospital. It is unlikely that in the presence of adequate medical resources, a casualty would be triaged as expectant.

4.1.2 Antidotes

The antidotes for nerve agent poisoning are **atropine**, which blocks the effects of the neurotransmitter or chemical that causes the overstimulation, and **2-PAMCI**, an oxime which removes the nerve agent from the enzyme. Ventilation and oxygen are necessary if the casualty is not breathing. An anticonvulsant, diazepam, might also be needed. The military and some other agencies have an autoinjector system for self use or 'buddy' use in treating cases of **confirmed** nerve agent poisoning. (During periods of high threat, autoinjectors may also be made available more widely to the emergency services in general.) However, it is important to understand that atropine:

◆ Is a poison (and a controlled drug)
◆ Unsuitable for use in all cases of nerve agent poisoning
◆ May prove lethal to a chemical casualty not suffering from nerve agent poisoning
◆ Is likely to cause serious side effects in some cases

In a normal adult **without** nerve agent poisoning, 2 mg of atropine will cause:

◆ An increase in heart rate of about 35 bpm (which can easily be tolerated in someone without heart disease)
◆ Mydriasis, blurred vision for about 24 hours, and drying of secretions including saliva and sweat. Inhibition of sweating can lead to heat storage problems if the ambient temperature is warm, if the

individual is doing any physical exertion, such as walking or if the casualty is wearing full PPE.

The recommended starting dose for infants and children under 2 years is 0.5 mg atropine and the recommended dose for children between 2 and 10 years is 1.0 mg atropine. It is recommended that wherever possible, the use of atropine is supervised by appropriately qualified medical personnel.

Healthy adults. The amount of pralidoxime chloride in 1 – 3 autoinjectors is easily tolerated by healthy adults. However, the pralidoxime by intravenous infusion will cause a marked prolonged hypertension if administered too rapidly. This should be given in not less than 20 minutes. Phentolamine (5 mg, IV) will transiently reverse the hypertension. The dose should not be repeated for an hour.

The recommended dose for healthy children is typically 15 mg/kg.

Typical symptoms and treatment
The following information is provided as guidance for trained medical personnel. It does not supersede any established medical practices or operational procedures.

4.1.2.1 Casualties with mild effects
Miosis alone: No antidotes are necessary usually. However, if eye/head pain or nausea and vomiting (in

the absence of other systemic signs suggesting a liquid exposure) are severe, atropine/homatropine eye drops should be administered (these will cause blurred vision).

Miosis and rhinorrhoea: Atropine only if rhinorrhoea is severe (2 mg is the standard dose).

4.1.2.2 Casualties with moderate effects
Shortness of breath: Atropine (2 mg for all but most severe cases) and 2-PAMCl (600 mg, if injector; 1 g in infusion over 20 to 30 minutes). Follow with additional atropine (2 mg) at 5 to 10 minute intervals until breathing is improved. Assisted ventilation and oxygen are rarely needed except in a casualty with cardiac or pulmonary disease.

Vomiting, diarrhoea (from liquid exposure): Same as for shortness of breath.

4.1.2.3 Severe effects
Convulsions, severe shortness of breath or apnea, unconsciousness, severe gastrointestinal effects, muscular twitching, or a combination of two or more of these: Atropine (6 mg, IM – not IV) and 2-PAMCl (1,800 mg – 3 injectors – if injectors are available; otherwise start infusion of 1 g to be given over 20 – 30 minutes); diazepam (5 – 10 mg slowly IV or ET, 10 mg IM). More atropine – 2 mg at 5 to10 minute intervals – until improvement is noted. 2-PAMCl should be given at hourly intervals for a total of 3 doses; it should be given as a slow

(20 – 30 minutes) infusion. Diazepam should be given to any severe casualty, convulsing or not. In a hospital, lorazepam is often used as an anticonvulsant instead of diazepam. Up to 20 mg of atropine may be needed.

Paediatric doses for diazepam are 0.2-0.4 mg/kg slow IV, ET or IM.

Atropine should not be given IV in a hypoxic patient; IV atropine regularly produces ventricular fibrillation in animals under these circumstances. The initial dose must be IM under this circumstance.

Atropine, unless administered in large doses, will not reverse miosis so pupil size should not be used as an indication of atropine effectiveness. Atropine eye drops will reverse the miosis, but cause blurred vision; these should be given only to relieve eye or head pain.

Atropine has very little effect on abnormalities in skeletal muscle movements.

The heart rate in nerve agent poisoning may be decreased, normal or increased. If the heart rate is very slow (<60 bpm) more atropine might be indicated, but a rapid heart rate does not necessarily indicate that enough atropine has been administered.

Note: Because of the intense airway resistance, ventilation will be unsuccessful until atropine reduces the bronchoconstriction. Therefore, if atropine is immediately available in the form of an autoinjector and can be administered within seconds, then atropine should be given before endotracheal intubation. However, since atropine and diazepam may both be administered via the endotracheal tube and since difficulties with intravenous cannulation may unnecessarily delay even limited life-saving ventilation, endotracheal intubation should be initiated simultaneously to attempting intravenous access and drug therapy. If only one action may be performed at a time (and autoinjectors are not available), then endotracheal intubation should precede intravenous administration of atropine.

4.1.3 Further care

Casualties in the first two categories, **mild** and **moderate**, will improve after one or two doses of atropine. Generally, those exposed to vapour need not be hospitalised after this improvement. Those exposed to liquid should be observed for 18 hours for further effects of the agent.

Casualties with severe effects may require ventilation for several hours after an uncomplicated exposure – that is, in the absence of hypoxic damage. Atropine should be repeated at intervals until secretions are dry or drying and until ventilation can be done with ease. 2-PAMCl should be repeated at hourly intervals for a total of 3 doses. Acidosis should be corrected, and the anticonvulsant diazepam should be administered.

4.2 Cyanide

◆ **Protect yourself** with the appropriate level of PPE
◆ **Remove casualty from contamination** and contamination from casualty
◆ **Get the casualty away from the source**, such as by moving him upwind or out of a contaminated building
◆ If it is absolutely certain that exposure was to vapour only, remove outer clothing. If there is a possibility of liquid contamination, all clothing must be removed and the casualty must be showered or washed with soap and water, dilute hypochlorite or water.
◆ ABCs (Airways, Breathing, Circulation)

4.2.1 Triage

Given the rapidity with which cyanide vapour/gas causes effects, by the time the responder arrives on the scene casualties will be asymptomatic, exhibiting acute effects, recovering from acute effects, or dead.

◆ **Expectant** – Those with no cardiac activity. Since it is rarely known how long the heart has been stopped, every effort should be made to resuscitate these casualties, if resources are available.
◆ **Immediate** – Those who exhibit acute effects (convulsions, apnoea). These must receive the antidotes and oxygen immediately if they are to survive.
◆ **Delayed** – Those who are recovering from severe acute effects (unconscious, but breathing).

Administration of the antidotes and oxygen will hasten their recovery.

◆ **Minimal** – Those who are asymptomatic more than a few minutes after possible exposure to cyanide vapour. These persons will remain asymptomatic and no antidotes or oxygen are needed.

4.2.2 Antidotes

Treatment for cyanide poisoning is very effective if administered in time. One drug used, a nitrite, causes the formation of a different form of haemoglobin in the red blood cell, and this form of haemoglobin 'pulls' the cyanide out of the cell so that the cell can return to normal. A second antidote, the sulphur compound thiosulfate, combines with cyanide in the blood to render it inactive. The administration of oxygen is also helpful. Acidosis should be corrected.

◆ Conscious, breathing casualty: No antidotes
◆ Unconscious, not breathing: **Amyl nitrite perle via bag ventilator (first aid, only until IV drugs can be given); sodium nitrite, 300 mg IV; sodium thiosulfate, 12.5 g IV. Half these doses might be administered later if there is no response to the first dose**.

Note: A cyanide kit is currently available that contains two ampules of 300 mg sodium nitrite in 10 ml of water, two ampules of 12.5 g sodium thiosulfate in 50 ml of water, 12 ampules of amyl nitrite inhalant, the

---

necessary materials for administration and instructions for use.

4.2.3 Further care
Oxygen with assisted ventilation; correct the metabolic acidosis; repeat sodium nitrite and sodium thiosulfate at 50 per cent the initial dose.

4.3 Blister agents (vesicants)
◆ **Protect yourself** with the appropriate level of PPE
◆ **Remove casualty from contamination** and contamination from casualty
◆ **Get the casualty away from the source**, such as by moving him upwind or out of a contaminated building
◆ If it is absolutely certain that exposure was to vapour only, remove outer clothing. (Remember, a blister agent could permeate through clothing yet have no effect on skin for several hours)
◆ If there is a possibility of liquid contamination, all clothing must be removed and the casualty must be showered or washed with soap and water, dilute hypochlorite or water

The vesicant mustard rarely causes immediate clinical effects. A casualty seen within an hour or two after exposure to this agent will have no signs or symptoms and will require no care other than decontamination and further observation. After inhalation of a very high concentration of vapour, airway effects may begin within

---

an hour or two. This type of casualty must be taken to an intensive care unit immediately. In this situation the chance of survival is slim. However, a concentration high enough to cause this would be very unlikely in most scenarios.

4.3.1 Triage

Mustard
Because the clinical effects of mustard do not appear until hours after contact with the agent, it is unlikely that someone responding to the scene will encounter a symptomatic casualty or a casualty to be triaged. In general, anyone potentially exposed to the agent should be classified as delayed and kept under observation.

◆ **Immediate** – airway effects within the first several hours after exposure to mustard suggest that the casualty received a large amount of agent and will have a severe, if not fatal, illness. This person should be triaged immediately for intensive pulmonary care.

◆ **Delayed** – likely to involve many casualties exhibiting effects from mustard. There is nothing to be done by the first responder for skin lesions, eye lesions or airway lesions. Contrary to procedures for eye damage from other causes, an eye injury from mustard does not make the casualty immediate. By the time the lesion appears, hours after exposure, mustard is gone from the eye. Symptomatic care is needed but will not reduce the damage.

◆ **Minimal** – a casualty with a small area of erythema or mild conjunctivitis that started more than 12 hours after contact with the agent

Lewisite and Phosgene Oxime

◆ **Immediate** – airway-effects within several hours of exposure suggest a large exposure and the need for immediate airway assistance. A person with eye effects should be decontaminated as soon as possible.

◆ **Delayed** – casualties with skin lesions without eye or airway damage. These casualties should be given pain medication, but no other immediate care is available.

◆ **Minimal** – a casualty with a very small skin lesion

4.3.2 Antidotes

Mustard

There is no specific antidote or treatment for mustard poisoning. The only initial action is to decontaminate as early as possible.

Lewisite and Phosgene Oxime

There is an antidote for Lewisite: **British-Anti-Lewisite**. If administered early enough, this will decrease some of the internal damage from Lewisite only. However, it will not help skin, airway or eye damage. Decontamination of Lewisite and Phosgene Oxime must be done immediately to prevent further tissue damage.

4.3.3 Further care

◆ *Skin:* Calamine or other soothing lotion for erythema. Some authorities feel that large blisters should be 'unroofed'; others feel that they should be left intact. Areas of denuded skin should be irrigated frequently (3 to 4 times a day), and this should be followed by the liberal application of a topical antibiotic.

Note: Fluid loss from vesicant burns is not of the same magnitude as fluid loss from thermal burns. Do NOT fluid load. Liberal use of systemic analgesics is recommended.

◆ *Eyes:* For mild lesions (that is, conjunctivitis), a soothing ophthalmic solution should be applied. For more severe lesions (for example, severe conjunctivitis, iritis, corneal oedema, lid inflammation and oedema): regular irrigation of the eyes followed by application of a topical antibiotic; regular application of a mydriatic (to prevent adhesions between lens and iris); and regular application of Vaseline to lid edges (to prevent adhesions). Some authorities believe that the use of topical steroids within the first 24 hours only will reduce inflammation. Topical analgesics should be avoided.

◆ *Airways:* For mild, upper-airway lesions: steam inhalation, cough suppressants. For more severe lesions: intubate early before laryngeal damage makes this difficult; assisted ventilation with oxygen; some authorities believe that early Positive End Expiratory Pressure (PEEP) ventilation will be useful; bronchodilators for evidence of bronchoconstriction; and antibiotics only after infecting organism is established.

4.4 Pulmonary agents

◆ **Protect yourself** with the appropriate level of PPE
◆ **Remove casualty from contamination** and contamination from casualty
◆ **Get the casualty away from the source**, such as by removal upwind or out of a contaminated building
◆ If it is absolutely certain that exposure was to vapour only, remove outer clothing
◆ If there is a possibility of liquid contamination, all clothing must be removed and the casualty must be showered or washed with soap and water, dilute hypochlorite or water
◆ ABCs: oxygen with or without assisted ventilation. Suction secretions.

4.4.1 Triage

Definitions
◆ **Immediate** – a casualty who is short of breath after exposure to a pulmonary agent
◆ **Delayed** – anyone possibly exposed to a pulmonary agent. These persons should be kept at absolute rest and under observation for at least 6 hours.

4.4.2 Antidotes
There is no antidote for pulmonary agents.

4.4.3 Further care
Assisted ventilation with oxygen. Intubate early before laryngeal spasm or oedema occurs.

PEEP may be of benefit. Bronchodilators if there is evidence of bronchoconstriction. Monitor fluid balance; administer colloids or crystalloids as indicated. Antibiotics are only appropriate after the organism is identified.

4.5 Riot control agents
♦ Eyes (impacted particle): Thoroughly flush the eyes to remove the particle. Removal of the particle by an ophthalmologist
♦ *Airways:* Ventilation with oxygen. Bronchodilators
♦ *Dermatitis:* Soothing lotions, such as calamine. Frequent irrigation. Prevent infection

4.6 Management of chemical casualties
A scene with multiple casualties is no place for an unprepared responder. This is especially true with an unknown toxic material. For example, some of the emergency responders in Matsumoto and in Tokyo did not fully appreciate the nature of the hazard and entered the incident scene without protection. Typically, these people became casualties themselves.

When the cause is unknown – as it usually is after the release of a toxic substance – **the only people who should enter the incident scene should be trained first responders equipped with the correct level of protective equipment.** If the initial responder does not know what happened and is not in PPE they report the incident to their control (using the appropriate format) and await assistance. Where possible, it is important that no unprotected persons be allowed to enter the scene.

Once a chemical release is suspected the scene will be cordoned and a major incident declared. The ultimate size of the cordon is likely to depend upon a number of interrelated factors, such as the amount of chemical released, the wind direction and speed, the temperature and so on. However, pending specialist advice, a cordon should be no less than 100 m from any point of known or suspected contamination

The area that is believed to be contaminated with a toxic material, liquid or vapour, is sometimes referred to as the 'Hot Zone'. The first responder in the appropriate level PPE (SCBA for those who intend to enter the scene) has several tasks. One of these is to start the egress of casualties out of the Hot Zone into a clean area, sometimes called the 'Cold Zone'. Ideally, the first stage is for the responder to move the casualties to an intermediate area or 'Warm Zone', where casualty decontamination takes place. These Zones do not automatically match the concept of inner and outer cordons. However, it is likely that the area within the inner cordon will include both the hot and warm zones.

The Hot Zone, Warm Zone and Cold Zone (clean area) should be clearly demarcated. In particular, the entrances and exits to each must be marked and subject to access control measures to prevent the movement of a potentially contaminated person from the Hot Zone to a cleaner area (For more information, see Site Set-up Procedures).

Casualty decontamination – on site considerations
Casualty decontamination is performed by responder personnel who are in the appropriate level of PPE. There is one decontamination area for those unable to walk (litter casualties) and another for ambulatory casualties. The type of decontamination, complete or partial, depends on the assessment of the situation and the casualty, as discussed below.

◆ The litter casualty station usually consists of several litter stands on to which the casualty is moved sequentially. On the first, clothing is removed – often by cutting if the casualty cannot assist. This must be done carefully so that the outer contaminated clothing does not come in contact with the skin or the inner or under clothing as it is removed.

◆ The next step is skin decontamination. This may be done on the same litter stand or on a second stand to which the casualty may be moved. Skin decontamination is done using a liquid to wash the agent off the skin.

◆ **Copious amounts of water may be used to remove the agent physically. Soap and water with a plain water rinse is also good for this purpose.** Another decontamination solution is hypochlorite, or household bleach. The military uses 0.5 per cent hypochlorite **(a dilution of 1 part of 5 per cent household bleach with 9 parts of water)** because 5 per cent hypochlorite will degrade the skin's ability to act as a barrier, and on a battlefield there is no way to rinse off the bleach. Although hypochlorite will cause the agent to decompose, this decomposition does not happen immediately and, as long as the agent-hypochlorite mixture is in contact with skin, some free intact agent will be present to damage or penetrate skin.

◆ The run-off from decontamination must be contained for proper disposal

◆ In the ambulatory casualty decontamination area the casualty is instructed to remove his clothing and enter the shower area. Soap should be provided and the casualty should be instructed to wash skin and all hair thoroughly.

◆ Any casualty who has had a possible exposure to a liquid chemical agent must be completely stripped and decontaminated with a liquid. This need not always be done for a casualty who has been exposed to a vapour only. Vapour might remain in clothing briefly if the casualty is not exposed to large amounts of fresh air with a breeze.

◆ To err on the side of safety, outer clothing should usually be removed. Vapour does not penetrate clothing, so there is little reason to remove inner clothing or underwear.

◆ Similarly, **vapour from nerve agents, cyanide and phosgene does not penetrate skin at concentrations likely to be present at the scene**. For example, five times the lethal concentration of sarin will cause no effects if the respiratory tract, nose and eyes are protected; a casualty exposed to this concentration will likely die from the exposure before reaching the decontamination area.

◆ Mustard vapour can damage skin but, by the time a casualty reaches the decontamination area, this will have dissipated either by evaporation or by absorption into the skin. However, the First World War experience suggests that mustard vapour might be in hair so, if mustard exposure is a possibility, these potential casualties must wash any exposed hair thoroughly.

After exiting the decontamination area, the casualty is provided with clothing or a covering and taken to the clean area, or 'Cold Zone'. Everyone exiting from the Hot Zone to the Cold Zone must pass through decontamination. Many in the Hot Zone will not be contaminated but, at this point, there is usually no way to distinguish them. Responders in PPE should shower with the PPE on and then carefully remove this equipment.

In the Cold Zone, more definitive triage is done, first aid is administered and casualties who need further care are sent to hospitals. In many circumstances, antidotes for chemical agent poisoning will not be available until the casualty reaches this area.

The principles of response to a chemical warfare agent incident are no different from the principles of response to an incident from an industrial chemical. In either case, the chemical may or may not be immediately identified. However, several factors associated with a terrorist attack are not commonly present in an industrial chemical spill:

◆ There may be secondary devices that are specifically designed to injure or kill responders
◆ The site of a terrorist attack is a crime scene and various police agencies, in particular, will need to enter the scene at a very early stage
◆ The material may be significantly more toxic than that encountered at HAZMAT accidents
◆ However, the same basic principles apply and the same procedures should be followed

Procedures at a chemical incident scene
◆ An unprotected person should not enter an incident area

◆ Initial responders to a chemical incident area should be suitably protected (*Note:* The fire service is most likely to be equipped to enter a chemical scene and recover casualties)

◆ Casualties should be sorted initially at the scene so that the most serious who may survive can be removed first

◆ Anyone exiting the contaminated area, including responders and bystanders must be decontaminated

CHAPTER 5: BIOLOGICAL AGENTS

Four components must be integrated in order to conduct an effective attack using a Biological Warfare (BW) agent:
(1) the **agent**
(2) the **munitions**
(3) the **delivery**
(4) the **meteorological conditions** at the target.

These four components must be considered irrespective of whether the threat is posed by a single individual, a well-equipped terrorist group (possibly state sponsored), or a country with a sophisticated BW programme.

5.1 Biological agent quick reference

	Inhalation Anthrax	Cholera	Pneumonic Plague
Likely method of dissemination	Spores in aerosol	1. Sabotage (food & water) 2. Aerosol	Aerosol
Transmissible man to man	No (except cutaneous)	Rare	High
Incubation period	1-7 (up to 43) days	12 hours - 6 days	1-3 days
Duration of illness	3-5 days (usually fatal)	> 1 week	1-6 days (usually fatal)
Lethality	High	Low with treatment, high with out	High unless treated within 12-24 hours
Vaccine efficacy (aerosol exposure)	2 doses of vaccine protects against 200-500 LD50s in monkeys	No data on aerosol	Vaccine no longer available

Source: USAMRIID, January 2002

	Tularemia	Q Fever	Ebola
Likely method of dissemination	Aerosol	1. Aerosol 2. Sabotage (food supply)	1. Direct contact (endemic) 2. Aerosol (BW)
Transmissible man to man	No	Rare	Moderate
Incubation period	1-21 days	10-40 days	4-16 days
Duration of illness	>2 weeks	Weeks	Death between 7-16 days
Lethality	Moderate if untreated	Very low	High for Zaire strain, moderate with Sudan
Vaccine efficacy (aerosol exposure)	80% protection against 1-10 ID50s	94% protection against 3,500 ID50s in guinea pigs	No vaccine

Source: USAMRIID, January 2002

	Smallpox	**Venezuelan Equine Encephalitis**	**Botulinum Toxin**
Likely method of dissemination	Aerosol	1. Aerosol 2. Infected vectors	1. Aerosol 2. Sabotage (food & water)
Transmissible man to man	High	Low	No
Incubation period	7-17 days	1-6 days	Variable (hours to days)
Duration of illness	4 weeks	Days to weeks	Death in 24-72 hours; lasts months if not lethal
Lethality	High to moderate	Low	High without respiratory support
Vaccine efficacy (aerosol exposure)	Vaccine protects against large doses in primates	TC-83 protects against 30-500 LD50s in hamsters	3 doses efficacy of 100% against 25-250 LD50s in primates

Source: USAMRIID, January 2002

	T-2 Mycotoxins	Ricin	Staphylo-coccal Enterotoxin B
Likely method of dissemination	1. Aerosol 2. Sabotage	1. Aerosol 2. Sabotage (food and water)	1. Aerosol 2. Sabotage (food supply)
Transmissible man to man	No	No	No
Incubation period	2-4 hours	Hours to days	1-6 hours
Duration of illness	Days to months	Days – death within 10-12 days for ingestion	Hours
Lethality	Moderate	High	<1%
Vaccine efficacy (aerosol exposure)	No vaccine	No vaccine	No vaccine

Source: USAMRIID, January 2002

5.2 General agent properties

The agent is the most important component. The **agent can consist of a toxin with a short incubation period whose effectiveness is, in general, limited to relatively small targets** (one or two square kilometres). Conversely, it could be a **bacterial or viral organism, which has a longer incubation period but which can cause casualties over wide areas** (hundreds of square kilometres). Depending on the agent selected, **casualties can be classified as:**

◆ **'incapacitating' (relatively few deaths)**
◆ **'lethal'**

There are several requirements a BW agent must meet. Though requirements can vary, typically an agent must be:

◆ infectious (toxic) to humans, crops and livestock
◆ capable of being produced in sufficient quantity to meet the target requirements
◆ stable during preparation, delivery to the target, dissemination and after release

Biological agents can be prepared and used either in liquid or dry form. Procedures and equipment for preparing liquid biological agents are simple, but the resulting product is difficult to disseminate in a consistent aerosol of the right particle size to guarantee injury. Conversely, procedures for producing dried biological agents are complex and require more sophisticated equipment, yet dried particles can be made of uniform size and can be readily disseminated by any number of crude devices. The terrorist is therefore faced with a choice between easily created but less effective aerosolised liquid agent and milled, dried agent which is costly and difficult to make.

Emergency and first responders should be aware of the physical characteristics of the BW agents they might encounter

5.2.1 Liquid agents
Liquid agents can be created by fermentation, tissue culture or from embryonated chicken eggs. These liquid materials can include bacteria, bacterial toxins, viruses or rickettsiae. They have common characteristics. **Liquid formulations of BW agents typically have a consistency similar to whole milk**.

The colour of liquid agents varies. **Bacterial agents derived from fermentation are generally an opaque amber to brown colour**. Egg-derived liquid agents will either retain **egg yolk's natural colour** since the entire egg contents have been processed, **or be slightly pink to red**, since only the embryo has been processed. BW toxins are similar.

5.2.2 Dried agents
Organisms with the capability to produce viruses by means of tissue culture may also be able to process the

liquid into a dried powder. **Dried agent will have the consistency of talcum powder.** An ideal dry agent should be free-flowing (free from lumps) with a consistent particle size of between 1 and 5 µm. Further sophistication would give the material a high electrostatic charge enabling it to cling to surfaces. Less sophisticated processing yields a coarse powder whose particle size ranges from 10 to 20 µm.

The colour of dried agent reflects the liquid from which it was derived:
◆ **dried bacterial agents tend to be amber to brown in colour**
◆ **viral agents derived from a tissue culture system will be an off-white**
◆ **viral and rickettsial agents derived from embryonated chicken eggs will be either brown to yellow or pink to red**

It should be noted that these characteristics could be camouflaged by the addition of dyes. For example, a light coloured powder which would be apparent immediately if spread across a road or pavement might be changed to a darker colour.

5.3 Biological delivery

5.3.1 Munitions
The BW munition has to disseminate the agent into a consistent aerosol to make it infectious in humans.

There is a wide variety of munition designs. Nations considering BW for overt, open-air warfare may develop sophisticated small bomblets that employ either explosive or gaseous energy to disseminate liquid or dried agents efficiently. These bomblets are usually released from missiles or high-performance aircraft and are referred to as:

1) 'line-source' munitions, and
2) 'point-source' munitions

The line-source munition is usually delivered from a high-performance aircraft with specially adapted drop-tanks or a cruise missile with a warhead containing the agent and the energetic release system. These munitions disseminate the agent at right angles to the prevailing wind direction, upwind of the target. They can infect much larger target areas than the point-source systems.

Terrorists would use much less sophisticated delivery systems such as garden or industrial sprayers. These devices, although readily available, are not designed to generate aerosols of consistently sized droplets. A well-organised terrorist group could develop a bomb to disseminate the agent. However, agents are delicate and explosive release will render a high proportion of the agent ineffective. Gas pressure generation, although the best way to generate high agent concentrations in aerosol form, is likely to be beyond the manufacturing capability of most terrorist groups.

The events of 2001 have shown that, although terrorists may not have the means to manufacture such materials, they may have the means to acquire them. The anthrax used in the October 2001 mailings was a sophisticated product with a very consistent particle size and mixed with Bentonite (a clay powder) to improve its flow properties and prevent clumping. Although the dissemination method was crude and erratic, it had the effect of causing great public fear and signified a perpetrator with considerable skill and access to top-quality production equipment and material.

Simple devices such as a suitably modified fire extinguisher, placed upwind of the intended target or at the air intake of a building, could potentially produce a large number of infections.

5.3.2 Delivery systems
Delivery systems for a BW agent can consist of just about anything that can produce an aerosol, including a wide variety of commercially available objects. Possible examples include:
◆ Crop spraying equipment (light aircraft or road vehicle mounted)
◆ Home garden sprayer
◆ A fire extinguisher
◆ Paint spraying equipment

BW munitions and delivery systems are very interdependent; frequently the munition dictates the delivery system.

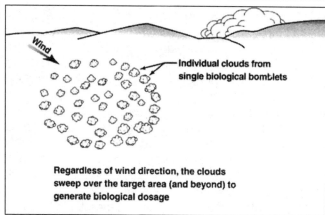

Point-Source Delivery 2002/0131357

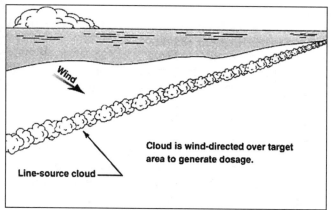

Line-Source Delivery 2002/0131356

CHAPTER 5: Biological Agents

◆ A **point-source delivery system**, using either sophisticated bomblets manufactured by a country or crude bombs prepared by a terrorist group, is illustrated in the diagram. The aerosol clouds from the individual bombs tend to merge and produce high levels of infection throughout the target.

◆ **Line-source delivery** can be achieved by either a sophisticated aircraft spraying a liquid or dry agent or by a terrorist walking or driving along a line while disseminating the agent from a home garden sprayer. The success of this type of delivery is highly dependent upon the prevailing meteorological conditions at the target. Line-source delivery is shown in the diagram.

5.4 Meteorological conditions

If BW agents are to be employed on targets in the open, the meteorological conditions at the target become extremely important. For an aerosol to be truly effective it must remain at 1 to 5 m above ground level. An inversion (a layer of cold air at the surface below a warmer layer above) is an ideal environment in which to disseminate an aerosol. Under these conditions, the cold air prevents the vertical mixing of air and the aerosol remains at ground level. **In still conditions, an inversion is most likely to occur between dusk and dawn, making it a higher risk period for a BW attack**. Sunlight (particularly the ultra-violet component) eventually destroys most BW bacterial and viral agents.

Exceptions are certain toxins and the spores of *Bacillus anthracis* and *Coxiella burnetii*.

◆ The wind is an important consideration in an open air BW attack. **Aerosols are most effective in windspeeds between 5 and 25 mph. At less than 5 mph, the aerosol's coverage will be limited. Above 25 mph, the aerosol will disperse too quickly and lose effectiveness.**

As a general rule, liquid agents perform best in humid environments. Dry agents perform best in dry environments. Extremes of temperature will adversely affect viability in both cases.

Clearly, a BW agent released in an enclosed space will be highly concentrated and therefore effective.

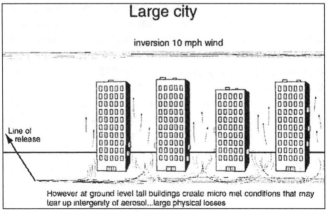

Biological weapons release in an urban environment **2002**/0131355

The urban environment, with its close concentration of tall buildings, can produce wind vortices which have unpredictable effects on agent dissemination.

5.5 Physics of the primary aerosol

After a munition disseminates the agent, there is a short period of time before the aerosol is fully formed. This period is known as aerosol equilibration. The larger particles (15 μm and larger) quickly fall to the ground. After about a minute, the aerosol comes to equilibrium with the atmosphere. The cloud will mainly comprise particles in the 1 to 5 μm range and it behaves as a gas. This phase is usually referred to as the "primary aerosol". A well-formed primary aerosol is critical to the success of a BW attack. Since it behaves as a gas, the primary aerosol can travel through heavily forested areas, for example, without degradation. **The aerosol permeates the atmosphere around an individual, and few, if any, infectious particles may stick to that person's clothing or other objects in the aerosol path**. At a normal breathing rate of 10 to 20 litres/min, the aerosol is quickly drawn into the lungs to cause infection. The working (breathing) rate of the individual will have a bearing on the concentration inhaled.

Contact exposure to the larger particles (which settled on horizontal surfaces) will cause a low risk of infection.

5.5.1 Secondary aerosols

Secondary aerosol generation is an added risk. In a high activity environment such as a battlefield or a busy city, the larger particles which settled after the attack can be re-aerosolised by machinery, passing vehicles, personnel or birds and animals. The risk from secondary aerosol infection is low.

Agent particles which settle tend to adhere strongly to the surface. However, powders can be treated to reduce or overcome this adhesive force and raise the infectivity of secondary aerosols. The dried agent powder requires special processing to achieve this. However, as the anthrax letters of October 2001 demonstrated, terrorists can obtain powdered agents with these properties, increasing the risk from secondary aerosolisation.

5.5.2 Biological decay in aerosols

Aerosol decay occurs through both physical decay (the fall-out of large particles) and biological decay (death of the biological agent). Next to respiratory virulence, biological decay is the most important factor in determining how far downwind an infectious or toxic aerosol can travel and still be effective. Biological decay is expressed in terms of percent death per minute of aerosol age and occurs in geometric progression. For example, a biological decay of 20 per cent/min implies that the total viable content of the aerosol is halved every 3.5 minutes of age.

5.6 Alternative BW delivery methods

5.6.1 Oral

The oral route, which involves the contamination of food and water supplies, can provide a less effective alternative pathway for dissemination of biological weapons. Contrary to media reports, **municipal water supplies are very difficult to contaminate effectively**. Dilution and diffusion, as well as chlorination and other treatments, render this route of attack almost ineffective. For example, reverse osmosis has proven effective against ricin, microcystin, T-2 and saxitoxin. Coagulation, flocculation and chlorination have also proved effective against certain types of toxins. However, **the targeted contamination of a specific water supply as it enters a building, for example, represents a feasible approach, with the right choice of agent**.

Individual foods and fruits can be contaminated at the point of manufacture and along the distribution pathway to the consumer. Here, the resulting infection can be acute, but limited to relatively small self-limiting populations (only clients of the selected supermarket chain, for example). Although a potent source of public anxiety, through sensational media attention, such attacks have usually been reserved for economic targets. The perpetrator has frequently been a disgruntled former employee or other grudge-holder.

5.6.2 Dermal exposure

Dermal exposure is not an effective means for the dissemination of biological weapons. **Intact skin provides an excellent barrier for most, but not all, biological agents**. However, mucous membranes, damaged skin, or open wounds constitute potential breaches of the natural dermal barrier through which biological agents may pass. The tricothecene toxins, and T-2 mycotoxin in particular, came to people's attention primarily because of reports in the 1970s of yellow (and other colour) powders, dust and 'rain' incidents in southeast Asia. People and domestic animals, in the open and exposed directly to the 'rain' reportedly became ill. Skin lesions were among the primary effects; these toxins are thought to penetrate the skin.

5.6.3 Vector transmissions

Disease can be spread by 'vectors', such as mosquitoes, ticks or fleas. These vectors can be produced in large numbers and can be contaminated by allowing them to feed on infected animals or infected blood reservoirs. While this is possible, the deliberate integration of two living systems as a biological weapon is difficult. However, the method is natural, low-tech and unlikely to arouse suspicion. It is perhaps typical of the type of unsophisticated technique popular with fundamentalist terrorists.

5.7 Bacteria

5.7.1 Anthrax

Bacteria: Anthrax		
Symptoms	**Incubation period**	**Treatment**
Cutaneous form: sores or blisters form on skin; Inhalation form: non-specific flu-like symptoms followed by respiratory distress, fever, shock, or death; Intestinal form: intense stomach pain, bowel obstruction, dehydration, . diarrhoea, fever, blood poisoning, death (rare in humans)	1 to 7 days (up to 43 days), most cases within 48 hours	Ciprofloxacin, doxycycline, penicillin

Anthrax is caused by the bacterium *Bacillus anthracis*. Anthrax bacteria are rod-shaped in microscopic form, gram-positive, have aerobic oxygen requirements and are sporulating. Recent cases have shown that the rate of anthrax incubation is dose dependent and can take as long as 43 days.

Symptoms and effects
Anthrax is an acute bacterial infection of the skin, lungs or gastrointestinal tract

◆ **Cutaneous anthrax** forms when the bacteria infect the skin. Such an infection causes dry scabs to form on the skin at the point of contact. **Traditionally**, this is caused by direct contact with contaminated wool, hides or tissues of infected herbivorous animals (cattle, sheep, goats). This form **- known as cutaneous anthrax**, malignant pustule and malignant carbuncle – occurs most frequently on the hands and forearms of persons who work with livestock. The swelling and ulcerated sores associated with cutaneous anthrax may develop into systemic infections, although **cutaneous anthrax is treatable with antibiotics**.

◆ **Inhalation anthrax** results from deposition of the bacterial spores in the lungs and causes fever, shock and eventually death. The disease begins with a sudden onset of flu-like symptoms within 1 to 7 days of exposure to the organism. After 2 to 4 days, victims exhibit a range of more severe symptoms (difficulty breathing, exhaustion, tachycardia, cyanosis and terminal shock). Death usually occurs within 24 hours of acute phase onset. **The fatality rate for untreated pulmonary anthrax is over 90 per cent.**

As a biological weapon, it is expected that anthrax spores would be released at a planned location to be inhaled. Iraqi and Soviet BW programmes worked to develop an anthrax weapon. While the exact reasons for the weaponisation of anthrax by these two countries are unknown, the bacteria may have been selected because of its special physical properties. **Its spores can survive sunlight for a few days and steam or heat up to 318°F (159°C), and can remain viable in soil or water for years or even decades.** About 8,000 to 10,000 spores are typically required to cause inhalation anthrax.

◆ Although the LD50 (the lethal dosage needed to kill at least 50 per cent of the persons within the target area) of anthrax is higher than some other biological agents, inhalation anthrax is much more lethal than common chemical agents. In 1970, the World Health Organization estimated that 125,000 people would become incapacitated, of which 95,000 would be expected to die, if an aerosolised form of anthrax were to be delivered over a city of 500,000.

◆ **Intestinal anthrax** in humans requires that a large number of spores be swallowed. Food poisoning with anthrax would bring on abdominal distress followed by fever, signs of septicaemia (presence of bacteria in the blood) and death.

5.7.2 Brucellosis

Bacteria: Brucellosis (undulant fever)

Symptoms	Incubation period	Treatment
Prolonged fever, headaches, profuse sweating, chills, pain in joints and muscles, fatigue	1 to 3 weeks, sometimes months	Doxycycline, plus rifampin

---

Brucellosis, also known as undulant fever, may be caused by infection with a number of Brucella species, notably *Brucella suis*, *Brucella abortus* and *Brucella melitensis*. These are gram negative rod-shaped, non-motile and non-sporulating bacilli.

Symptoms and effects
Likely methods of dissemination in a BW context would be aerosol or sabotage of a food supply. The incubation period typically ranges from 5 to 21 days, but occasionally lasts several months. **The bacterium causes incapacitation rather than fatality in most cases**, but the fatality rate averages around 2 per cent (between a range of 2 to 13 per cent). **Symptoms may include intermittent fever, headaches, sweating, chills, malaise, body aches and anorexia.**

5.7.3 Cholera

Bacteria: Cholera		
Symptoms	**Incubation period**	**Treatment**
Acute infectious gastrointestinal disease, vomiting, diarrhoea, rapid loss of fluids, severe muscular cramps, collapse	Hours to 5 days, usually 2 to 3 days	Oral rehydration therapy, tetracycline

Cholera is caused by infection with *vibrio cholerae*. In microscopic form it appears as a slightly bent rod. The bacterium is motile, gram-negative and non-sporulating.

Symptoms and effects
Cholera is an acute human gastrointestinal disease, characterised by sudden onset of nausea, vomiting, profuse watery diarrhoea with rice water appearance, rapid loss of bodily fluids, toxemia and collapse.

The primary route of exposure is through contact with faecal matter or contact with contaminated water. The incubation period can last from 1 to 5 days, with 3 days being the average.

Vibrio cholerae is readily killed by dry heat at 243°F (117°C), by steam and boiling, by short exposure to disinfectant, and by chlorination of water. It can thrive in saline water or water polluted with organic matter for up to 6 weeks. *Vibrio cholerae* is killed by standard water treatment methods. Cholera is fatal in 3 to 30 per cent of treated cases and 50 to 80 per cent of untreated cases. The organism is not appropriate for aerosol delivery.

5.7.4 Glanders

Glanders is an infection caused by the gram-negative rod, *Burkholderia mallei* (formerly known as *Pseudomonas mallei*).

Symptoms and effects

Glanders usually infects equids (horses, asses and mules). Carnivores can be infected if they eat the meat of animals killed by the disease. Rarely, goats and sheep can be infected. However, **humans can contract the disease from infected animals, since it enters through the mouth or nose, affecting nasal and respiratory tissues**. It is potentially a highly lethal disease and clinical response to most modern antibiotics remains untested. Although quite rare in humans, almost all untreated cases of systemic glanders are fatal and lethality for chronic cases can range from 50 to 70 per cent. The disease, in humans, can have an incubation period ranging from a few days to several weeks. **The disease can be transmitted via small particle aerosol or through skin abrasions, causing large lesions, and ulcers in skin, mucous membrane and visceral tissues. In chronic glanders abscesses can affect all tissues of the body.**

5.7.5 Melioidosis

Melioidosis is caused by infection with *Burkholderia pseudomallei*. The bacteria are non-sporulating, gram-negative, rod-shaped in microscopic form, motile and aerobic.

Symptoms and effects

Melioidosis **is a widely distributed environmental bacteria of moist environments that can be contracted through breaks in the skin, ingestion of contaminated food or water, or via inhaled, contaminated aerosols**.

Clinical disease is not common in the developed world. It may simulate typhoid fever or tuberculosis including **pulmonary cavitation, empyema, chronic abscesses and osteomyelitis**. In the absence of antibiotic therapy, the disease is nearly always fatal. After only a few days of incubation, **the acute cases can be accompanied by a sudden onset of chills, headache, fever, muscle/joint pain, rapid prostration (exhaustion), a cough or hard breathing, nausea and vomiting**. Death is rapid, coming in as little as 10 days.

The disease may be attractive as a biological weapon precisely because it is a rare disease outside the tropics, and there currently is no vaccination against it.

5.7.6 Plague

Bacteria: Plague		
Symptoms	**Incubation period**	**Treatment**
High fever, headache, general aches, extreme weakness, glandular swelling, pneumonia, haemorrhages in skin and mucous membranes possible, extreme lymph node pain	Bubonic: 2 to 6 days, unvaccinated; few days longer, vaccinated Pneumonic: 1 to 6 days	Streptomycin. Doxycycline, gentamicin and ciprofloxacin are alternates. chloramphenicol for Plague meningitis

Plague is a disease caused by infection with the bacterium *Yersinia pestis* (formerly known as *pasturella pestis*). The organism is aerobic, rod-shaped, gram-negative, non-sporulating and non-motile.

Symptoms and effects

Plague occurs in two primary clinical types in humans. Plague, or the "Black Death", **is transmissible to humans by the bite of an infected flea, or from person to person by the respiratory route. In general, it is characterised by a rapid clinical course with high fever, extreme weakness, glandular swelling and pneumonia**. If untreated the disease steadily progresses until the victim dies. Haemorrhages in the skin and mucous membrane may or may not occur. The disease only lasts for 1 to 2 days before death occurs (if untreated).

◆ **Bubonic plague**, the most common type of plague, is transmitted from rodents to humans by the bite of an infected flea; the disease in nature is then perpetuated by a rodent-flea-rodent transmission cycle. The bacilli spread through the lymphatic system, causing enlarged lymph nodes (buboes) in the groin. The bacilli escape from the nodes, invade the bloodstream and produce a generalised (septicaemic) – often fatal – infection. The spleen, lungs and meninges may be affected.

◆ **Pneumonic plague, which may result from the septicaemic form or from inhalation of the organism, spreads rapidly until haemorrhagic pneumonia involves the entire lung area**. Untreated pneumonic plague is usually fatal. Plague has been considered an attractive agent for biological warfare since it is extremely infective, especially in the pneumonic form. Disseminating plague in aerosolised liquid droplets ranging from one to five µm in diameter could prove extremely lethal against unprotected, unwarned populations. Historically, difficulties in sustaining the organism's virulence and its lack of stability have been obstacles to weaponisation of this bacterium.

5.7.7 Tularemia, Rabbit fever or Deer fly fever

Tularemia is caused by *Francisella tularensis*. Varying in size and shape, the tularensis bacteria are small, non-sporulating, non-motile, aerobic, gram-negative cocci-bacilli.

Symptoms and effects

Tularemia occurs in a variety of clinical forms, depending upon the route of inoculation and virulence of the strain. Therefore, the lethality rates for the disease depend on the characteristics of a particular strain.

Humans acquire the disease under natural conditions through inoculation (ulceroglandular form) of skin and mucous membranes with blood or tissue fluids of infected animals, or bites of infected deer flies, mosquitoes or ticks. Less commonly, inhalation of contaminated dust or ingestion of contaminated foods or water may produce clinical disease.

◆ **Ulceroglandular tularemia begins with an ulcer at the site of inoculation. Lymph nodes proximal to this inoculation site are typically swollen and tender and systemic symptoms such as fever and malaise typically follow.**

◆ **Typhoidal tularemia manifests as fever, prostration (fatigue or extreme tiredness) and weight loss, but without adenopathy (swelling**

Bacteria: Tularemia (Rabbit fever or Deer fly fever)

Symptoms	Incubation period	Treatment
Inhalational tularemia is characterised by sudden onset of chills, fever, headache, muscle aches, fatigue, loss of body fluids possibly accompanied by typhoid-like symptoms (most serious form); skin infections are characterised by deep ulcers found on skin with swelling of regional lymph nodes	1-21 days, average being 3 days	Streptomycin or gentamicin or doxycycline, chloramphenicol, ciprofloxacin

---

of the lymph glands). Diagnosis of primary typhoidal tularemia is difficult, as signs and symptoms are non-specific and there is frequently no suggestive exposure history.

Francisella tularensis is very infective, and a single bacterium has the potential to infect a human and cause symptoms ranging from chills and fever, to pneumonic problems and severe prostration. The infectivity rate is 90 to 100 per cent, meaning that most unprotected people would demonstrate symptoms of the disease in a biological weapons attack. A BW attack with tularensis would most likely be delivered by aerosol, thereby causing a severe pneumonia.

5.7.8 Typhoid fever

Bacteria: Typhoid fever		
Symptoms	**Incubation period**	**Treatment**
Dull frontal headache, fever, rose coloured spots on skin, constipation or diarrhoea, abdominal tenderness	3 to 60 days depending on dose, usually 7 to 14 days	Typhoid vaccine, antibiotics

5.8 Rickettsia

Rickettsia: Typhus (endemic or epidemic)		
Symptoms	**Incubation period**	**Treatment**
Headache, high fever, general aches and pains, chills, rash	6 to 14 days, usually 12 days	Tetracycline, chloramphenicol, supportive treatment needed

Typhoid fever is caused by infection with the Gram negative bacterium, *Salmonella typhi*. It is a rod-shaped, motile, non-sporulating organism.

Symptoms and effects

Typhoid fever is an infection that affects the entire body. **It is characterised by prolonged fever, lymphoid tissue involvement, ulceration of the intestines, enlargement of the spleen, rose coloured spots on the skin and constipation or diarrhoea.** Although the disease is lethal only in limited instances of untreated cases (usually less than 10 per cent lethality), it is very incapacitating. Normal incubation periods last from 7 to 14 days, and **the disease can be transferred through direct or indirect contact with contaminated substances or infected humans**. The bacteria can stay alive in water for 2 to 3 weeks, faeces for 1 to 2 months and snow and ice for 3 months.

Within the context of BW, *Salmonella typhi* would not be likely employed via aerosol dissemination. The organism lends itself to covert use by the contamination of food and limited volumes of water. The oral ID50 dose is estimated to be approximately 10,000,000 organisms.

5.8.1 Epidemic typhus/Endemic typhus
Epidemic typhus is caused by *Rickettsia typhi*. Endemic typhus is caused by *Rickettsia prowazekii*.

Symptoms and effects
Epidemic typhus is an infectious disease of humans that has a short and severe course. **It is characterised by severe headaches, sustained high fever, generalised muscle aches and a skin rash.** The route of exposure is through contact with **infected human body lice or their faeces**. The incubation period may last from 6 to 15 days.

Endemic typhus is similar to classic epidemic typhus except that the disease **is milder and has a somewhat slower onset.** The routes of exposure for endemic typhus are rodent flea bites, and exposure to contaminated rodent faeces. The incubation period is between 6 and 14 days.

5.8.2 Q fever

Rickettsia: Q fever		
Symptoms	**Incubation period**	**Treatment**
Sudden onset of fever, headache, chills, weakness, profuse perspiration, upper respiratory problems, mild coughing; chest, muscle, and joint pain	2 to 3 weeks depending on infecting dose	Tetracycline and ciprofloxacin; supportive treatment needed

Q fever is caused by *Coxiella burnetii.*

Symptoms and effects
Q fever is a zoonotic disease caused by the rickettsia *Coxiella burnetii*. The most common animal reservoirs are sheep, cattle and goats. Humans acquire the disease by inhaling particles contaminated with the organisms. A BW attack would cause disease similar to that occurring naturally. Rickettsia is highly infectious when delivered as an 'inhalable' cloud, which is the likely method of BW attack.

Q fever is an incapacitating agent and would not be expected to cause significant fatalities. In addition to its ability to infect and incapacitate, Q fever is one of the hardiest biological agents, making production and storage very easy. The rickettsia reproduces quickly in chick embryos and when dried has been found to hold 20 trillion organisms per gram. **The rickettsia are very stable as an aerosol in temperatures ranging from –126°F to 104°F (–52°C to 40°C). Like sporulating bacteria, extremely dry conditions do not harm these organisms, which can live on surfaces from 5 to 60 days.**

Studies have shown that one inhaled *C. burnetii* organism can infect a human and even **cause mild complications such as chills, headache, fever, chest pains, weakness, perspiration or loss of appetite.** Although Q fever's incubation period is longer than most biological agents, ranging from 10 to 14 days, large doses of Q fever can expedite the onset of symptoms. In terms of differential diagnosis, Q fever usually presents as an undifferentiated febrile illness (a fever of unknown origin), or a primary atypical pneumonia, which must be differentiated from pneumonia caused by mycoplasma, legionnaire's disease, psittacosis or chlamydia pneumoniae. More rapidly progressive forms of pneumonia may appear as bacterial pneumonias, including tularemia or plague. Staining sputum or serological confirmation provides the specific laboratory diagnosis.

5.9 Toxins

5.9.1 Botulinum toxin

Neurotoxins: Botulinum			
Natural source	**Rate of action**	**Stability/ persistence**	**Usual route of entry**
Bacteria (Clostridium Botulinum)	1-12 hours	Stable but non-persistent; stable 7 days in untreated water, 12 hours in air; destroyed by bases or boiling 15 minutes	Ingestion (contaminated foods) [Inhalation (BW attack)]

Botulinum toxin is produced by *Clostridium botulinum*, which is an anaerobic, partially motile, sporulating rod. Growth and toxic formation occurs under anaerobic conditions, such as in non-acidic meat samples, inside vegetable cans and in soil. While the bacteria reproduce by anaerobic fermentation, the toxin does not self-reproduce or multiply once inside the body. Botulism poisoning happens because of the pathogenicity of the already-produced toxin.

It is one of the most lethal toxic compounds known to man by the oral and intraperitoneal (intra-abdominal) routes. However, its toxicity diminishes significantly when inhaled.

Symptoms and effects

In its natural form, botulinum toxin is most often found in improperly canned or undercooked foods containing the *Clostridium botulinum* bacteria, which are heat resistant. The bacteria themselves are harmless. **Botulinum toxin is not contagious but is highly toxic** and most humans are susceptible to its powerful effects. Without prophylactic or antitoxin treatments, untreated victims may stand little chance of survival once intoxicated. If untreated, botulism is fatal in 60 per cent of cases because of muscle paralysis and respiratory failure. Symptoms may progress from as little as 24 hours from onset, but other studies indicate that symptoms may be delayed if the toxin is inhaled rather than ingested.

There are at least seven recognised types of the toxin (A, B, C, D, E, F and G), based on the antigenic specificity of the toxin produced by each strain. All produce similar symptoms. Botulinum toxin causes flaccid paralysis by blocking motor nerve terminals at the myoneural junction. **The first symptoms of paralysis are drooping eyelids, dry mouth and throat, difficulty talking and swallowing, and blurred and double vision. From the face, the flaccid paralysis progresses symmetrically downward to the throat, chest and extremities. When the**

diaphragm and chest muscles become fully involved, breathing becomes difficult and death from asphyxia results.

A high rate of lethality, a staunch resistance to treatments and a rapid onset of severe symptoms make botulinum toxin a suitable agent for biological warfare, particularly by oral administration. The human respiratory dose for botulinum toxin (A) is not known; however, in the Rhesus monkey, the aerosol dose is about 200 mouse intraperitoneal LD^{50} doses. If the Rhesus LD^{50} is extrapolated to a 155 lb (70 kg) man, on a weight-by-weight basis, the total dose required for man is at least 14,000 mouse intraperitoneal doses or 4.88 µm (with 50 per cent pure toxin). This dosage tends to reduce its effectiveness as a BW agent when delivered by aerosol. However, deaths from the ingestion of botulinum toxin in normal circumstances produce fatality rates of at least 60 per cent.

5.9.2 Ricin

Ricin is a toxin made from the mash that is left over after processing castor beans for oil. Castor bean processing is a worldwide activity; therefore, the raw materials for making ricin are easily available. Ricin is easy to produce and is stable.

Ricin was used in 1978 by Bulgarian intelligence operatives in the 'umbrella murder' of Georgii Markov, a Bulgarian dissident. A ricin-tipped 'bullet' was

Cytotoxins: Ricin

Natural source	Rate of action	Stability/ persistence	Usual route of entry
Castor bean seeds (*Ricinus communis*)	1-12 hours/ 5 min-1 hour	Very stable; stable in water or dilute acids; persistent	Inhalation/ ingestion

discharged into the victim and he died a day after the attack. Ricin, being a chemical compound, is also included on the prohibited Schedule I chemicals list of the Chemical Weapons Convention.

Symptoms and effects

In a BW scenario, it is expected that ricin would be released as a toxic cloud. It could also be injected into specific persons as a terrorist or sabotage weapon. The toxic effects of ricin occur because it kills the body cells that it contacts when it is taken into the body.

Though human experience with ricin is scarce, it is possible to predict symptoms that would result from contact, depending on whether ricin was inhaled, ingested or injected. **About three hours after inhaling ricin, the likely symptoms are coughing, tightness of the chest, difficulty breathing, nausea and**

muscle aches. This would progress to a severe inflammation of the lungs and airways, increased difficulty breathing, cyanosis (blue skin), and death within 36 to 48 hours from failure of the breathing and circulatory systems. The estimated aerosol dose for man is projected to be high (at least 320 mg) and therefore it would be disseminated from chemical agent munitions. **Ingestion of ricin would be expected to cause nausea and vomiting, internal bleeding of the stomach and intestines, failure of the liver, spleen and kidneys, and death by collapse of the circulatory vessels**. If injected, ricin causes marked death of muscles and lymph nodes near the site of injection and probable failure of major organs, leading to the death of the individual.

5.9.3 Saxitoxin

Neurotoxins: Saxitoxin			
Natural source	**Rate of action**	**Stability/ persistence**	**Usual route of entry**
Contaminated shellfish (dinoflagellates of Gonyaulax)	5 min-1 hour	Stable to heat and acids; sensitive to alkali; relatively persistent	Ingestion (contaminated shellfish)

Saxitoxin is the parent compound of a family of chemically related neurotoxins. In nature they are predominantly produced by marine dinoflagellates, although they have also been identified in association with organisms such as blue-green algae, crabs and the blue-ringed octopus.

Saxitoxin and its derivatives are water-soluble compounds that bind to the voltage-sensitive sodium channel, blocking propagation of nerve-muscle action potentials.

Symptoms and effects

Human intoxications are principally due to the ingestion of bivalve molluscs, which have accumulated dinoflagellates during filter feeding. The resulting intoxication, known as Paralytic Shellfish Poisoning (PSP), is known throughout the world as a severe, life-threatening illness. After oral exposure, absorption of toxins from the gastrointestinal tract is rapid. **Onset of symptoms typically begins 10 to 60 minutes after exposure, but may be delayed several hours depending on the amount ingested and other factors. Initial symptoms include numbness or tingling of the lips, tongue and fingertips followed by numbness of the neck and extremities, and general lack of muscular co-ordination.** Other symptoms may include a feeling of light-headedness or floating, dizziness, weakness, aphasia, incoherence, visual disturbances, memory loss and headaches.

Cranial nerves are often involved, especially those responsible for eye movements, speech and swallowing. **Respiratory distress and flaccid muscular paralysis are the terminal stages and can occur 2 to 12 hours after intoxication**. Death results from respiratory paralysis. In a BW scenario the most likely route of delivery is by inhalation or toxic projectile.

5.9.4 Staphylococcus Enterotoxin B (SEB)

Cytotoxins: Staphylococcus enterotoxin (B)			
Natural source	**Rate of action**	**Stability/ persistence**	**Usual route of entry**
Bacteria (Staphylo- coccus aureus)	1-12 hours usually 3-4 hours	Stable in heat, acids, alkali; relatively non- persistent	Ingestion (food poisoning)

Staphylococcus aureus produces many different types of toxin as it metabolises its growth substrates. The toxin that is considered to be a prime BW candidate is *Staphylococcus enterotoxin B (SEB)*. This toxin has inherent properties such that it could be employed either on open-air targets by aerosol dissemination or used to contaminate food and water. (It is a common cause of naturally acquired food poisoning in humans.) SEB is a very stable toxin and requires a very small dose (ED_{50} = 0.025 mcg total) to intoxicate humans by the respiratory route.

Symptoms and effects
The symptoms of intoxication via the inhalation route appear 1 to 6 hours following exposure and include fever, chills, headache, myalgia and non-productive cough.

Fever may read 102 to 109°F and last for 2 to 5 days. The cough may persist for 1 to 4 weeks. Many patients also suffer nausea, vomiting and diarrhoea. Thus, aerosol delivery of the toxin produces a wider range of more severe symptoms than those caused by ingestion where symptoms are typically limited to vomiting and occasionally diarrhoea. Even so, the mortality rate remains low – less than 2 per cent.

SEB toxin could be used to effectively contaminate food and limited water supplies in a variety of different scenarios. Salad bars in restaurants would be a good example. Since the effective oral dose for humans is quite low, 0.0004 μg/kg, very small amounts of toxin could be used and therefore the contaminated food would not be detected.

5.9.5 Trichothecene mycotoxins

Cytotoxins: Tricothecene mycotoxins			
Natural source	**Rate of action**	**Stability/ persistence**	**Usual route of entry**
Moulds on infected grains (Fusarium)	5 min-1 hour	Very stable; resists heat, acids; persistent	Ingestion/ inhalation/ absorption

Trichothecene mycotoxins are low molecular weight, non-volatile compounds produced by filamentous fungi (moulds) of the genera *Fusarium, Myrotecium, Trichoderma, Stachybotrys* and others. The structures of almost 150 trichothecene derivatives have been identified. Mycotoxins allegedly were used in aerosol form ('yellow rain') to produce lethal and non-lethal casualties in Laos (1975-1981), Kampuchea (1979-1981) and Afghanistan (1979-1981). It is suspected that there were more than 6,300 deaths in Laos, 1,000 in Kampuchea and 3,000 in Afghanistan. The alleged victims were usually unprotected civilians or guerrilla forces. These attacks often were alleged to have occurred in remote jungle areas, which made confirmation of attacks and recovery of agent samples extremely difficult.

Symptoms and effects

T2 and other mycotoxins may enter the body through the skin and aerodigestive epithelium (mouth, throat, lungs and gastrointestinal system). They are potent inhibitors of protein and nucleic acid synthesis. Their main effects are on bone marrow, skin, mucosal epithelia and germ cells. In a successful biological weapons attack with T2, the toxin will adhere to and penetrate the skin, or be inhaled or swallowed. The tricothecene toxins are not particularly toxic by the respiratory route and appear to be more effective by the fall-out of large particles from primary aerosols that land on the skin. The human respiratory dose is estimated to be 25 to 50 mg/kg of body weight, whereas the dermal route requires only 2.4 to 8 mg/kg of body weight.

Early symptoms begin within minutes of exposures and include burning skin, redness, tenderness, blistering and progression to skin necrosis with leathery blackening and sloughing of large areas of skin in lethal cases. Nasal contact is manifested by nasal itching and pain, sneezing, epistaxis (nosebleeds) and rhinorrhoea (watery discharge from the nose). Pulmonary/trachebronchial exposure is characterised by dyspnoea (difficulty breathing), wheezing and coughing. Mouth and throat exposure is characterised by pain and blood tinged with saliva and sputum. Anorexia, nausea, vomiting and watery or bloody diarrhoea with abdominal cramps occur with gastrointestinal

toxicity. Eye pain, tearing, redness, foreign body sensation and blurred vision may follow exposure to the eyes. Systemic toxicity is manifested by weakness, prostration, dizziness, ataxia (inco-ordination of the gait) and an overall loss of co-ordination. Tachycardia (rapid heart beat), hypothermia (low body temperature), and hypotension (low blood pressure) follow in fatal cases. **Death may occur in minutes, hours or days.**

5.10 Viruses

5.10.1 Chikungunya virus
Chikungunya is Swahili for "that which bends up", in reference to the stooped posture of patients afflicted with severe joint pain associated with the disease. The virus has been isolated from humans and mosquitoes in eastern, southern, western and central Africa and in southeast Asia.

Symptoms and effects
The virus is communicated to humans from bites of the *Aedes aegypti* mosquitoes, and epidemics are sustained by human-mosquito-human transmission. The incubation period for this virus is from 3 to 12 days. Once established, chikungunya runs a 3 to 7 day course, followed by an extended convalescence. Its effects are incapacitating and it has less than a 1 per cent fatality rate. **Symptoms may include arthralgic-exanthemic syndrome plus chills, fever, headache, nausea and vomiting**. The virus has a sudden onset,

sometimes biphasic. Similarities between it and dengue fever account for misclassification and under-reporting of chikungunya fever in areas with endemic dengue; therefore, laboratory confirmation of reported cases is important.

Chikungunya virus grows to a high concentration when grown correctly; moreover, the virus is quite stable in various environmental conditions. Since it can cause infections in primates via the aerosol route, this virus has properties that make it a likely candidate for biological warfare.

5.10.2 Congo-Crimean haemorrhagic fever virus
This virus is incapacitating-lethal, and has a 15 per cent to 50 per cent fatality rate. **Symptoms may include haemorrhagic syndrome plus nausea, vomiting, enlarged liver and occasionally prostration/coma**. The disease has a sudden onset, and haemorrhagic symptoms begin on or about the fourth day. The disease has a 2 to 12 day incubation period and a 9 to 12 day duration.

5.10.3 Dengue fever
Dengue is a mosquito-borne infection that in recent years has become a major international public health concern. Dengue is found in tropical regions around the world, predominantly in urban and peri-urban areas. The spread of dengue is attributed to expanding geographic distribution of the four dengue viruses and

of their mosquito vectors. The disease is now endemic in more than 100 countries in Africa, the Americas, the eastern Mediterranean, southeast Asia and the western Pacific. There are four distinct viruses that cause

---

Viral: Dengue fever

Symptoms	Incubation period	Treatment
Grades of severity, haemorrhagic fever, chills, intense headache, backache, excruciating joint and muscular pain, weakness, prostration, irregular rash, loss of appetite and constipation, abdominal discomfort with colicky pains and tenderness, spontaneous bleeding into the skin, gums, and gastrointestinal tract, circulatory failure, profound shock.	3 to 15 days, usually 5 to 8 days	Supportive care essential, no specific therapy, avoid aspirin, may need transfusions

---

dengue, and infection by one does not offer protection against subsequent infection by the other three.

Symptoms and effects

This severe, flu-like illness affects infants, young children and adults but rarely causes fatalities. The clinical features of dengue fever vary according to the age of the patient. Infants and young children may have an undifferentiated febrile disease with rash. Older children and adults may have either a mild febrile syndrome or the classical incapacitating disease **with abrupt onset and high fever, severe headache, pain behind the eyes, muscle and joint pains and rash**. **Dengue haemorrhagic fever is considered by most experts to have a low priority for biological warfare**.

5.10.4 Ebola virus

Ebola virus is one of the most pathogenic viruses known to science, causing death in 50 to 90 per cent of all clinically ill cases.

Symptoms and effects

The Ebola virus is transmitted through direct contact with the blood, secretions, organs or semen of infected persons. Health care workers have proved especially vulnerable. The virus has an incubation period of 2 to 21 days, and is **characterised by sudden onset of fever, weakness, muscle pain, headache and sore throat. This is followed by vomiting, diarrhoea, rash, limited kidney and liver functions, and both internal**

and external bleeding. The haemorrhagic symptoms begin on about the fifth day.

Ebola virus must be considered to have a high priority as a candidate for biological warfare. Moreover, this virus was assigned a high priority in the offensive BW programme of the former Soviet Union.

5.10.5 The Marburg virus

Marburg virus, which is caused by a virus closely related to the Ebola virus, was first recognised during an outbreak of a severe haemorrhagic disease associated with the importation of African green monkeys from east Africa to Germany. Subsequently, isolated human cases have been reported, primarily from sub-Saharan Africa, but virtually nothing is known about the epidemiology of the disease.

Symptoms and effects
After a 5 to 7 day incubation period, the virus produces headaches, fever, muscle pain, vomiting, diarrhoea, haemorrhagic diathesis, conjunctivitis, photophobia, skin rash and jaundice.

Fatalities occur in about 25 per cent of cases.

Marburg must be considered to have a high priority as a BW agent. **It is highly infectious by the aerosol route** and was actually weaponised in the former Soviet BW programme.

5.10.6 Junin virus
Also known as Argentine haemorrhagic fever.

Symptoms and effects
The virus has a 7 to 16 day incubation period, a 16-day duration, and a two-week convalescence. The disease has a gradual onset, with **symptoms that may include haemorrhagic syndrome plus chills, sweating, exhaustion and stupor**. Haemorrhagic symptoms occur within 5 days in more severe cases, and the virus is incapacitating-lethal with an 18 per cent fatality rate.

5.10.7 Rift Valley Fever

Viral: Rift Valley Fever		
Symptoms	**Incubation period**	**Treatment**
Headache, general aches and pains, nausea, vomiting, photophobia	2 to 6 days	Supportive care, no specific therapy

Rift Valley Fever (RVF) is a viral disease caused by the RVF virus. The virus circulates in sub-Saharan Africa as a mosquito-borne agent. Epizootics occur when susceptible domestic animals are infected and, because of the large amount of virus in their serum, amplify infection to biting arthropods. Deaths and abortions among susceptible species such as cattle and sheep provide a key diagnostic clue and method of surveillance.

Symptoms and effects
Humans become infected by the bite of mosquitoes or by exposure to virus-laden aerosols or droplets.

Although the disease may occur during a typical rainy season, outbreaks are typically associated with very high densities of arthropod vector populations that may occur during heavy and prolonged rains or in association with irrigation projects. In a BW attack, the disease would most likely be disseminated by aerosol, which would bring on a characteristic set of symptoms in humans and cause disease in sheep and cattle in the exposed area.

The incubation period for RVF is 2 to 5 days and is usually followed by an incapacitating febrile illness of similar duration **that features fever, conjunctival infection and sometimes abdominal tenderness.** A few petechiae (speckled haemorrhagic areas on the skin) or epistaxis may occur. A small proportion of cases (approximately 1 per cent) will progress to a viral haemorrhagic fever syndrome, often with associated hepatitis. These cases may manifest petechiae, mucosal bleeding, icterus (jaundice or yellowness of the sclera), anuria (lack of urine output) and shock; mortality in this group may be up to 50 per cent. A similar proportion will develop significant changes in the eye, including macular lesions associated with retinal vasculitis, haemorrhage, oedema and infarction (loss of blood supply). These begin after the patient begins convalescence from acute illness and about half of the patients will have permanent visual defects.

5.10.8 Smallpox

Viral: Smallpox		
Symptoms	**Incubation period**	**Treatment**
Severe fever, small blisters on the skin, bleeding of skin and mucous membranes	4 to 17 days, but typically with 10 to 12 days before onset of illness, 2 to 4 days more to onset of rash	No specific therapy, supportive care must include prevention of secondary infections

Smallpox is caused by the variola virus. Endemic smallpox was declared eradicated in 1980 by the World Health Organization (WHO). Although two WHO-approved repositories of variola virus remain at the CDC in Atlanta, Georgia, and at Vector in Koltsovo, Russia, the extent of clandestine stockpiles in other parts of the world remains unknown. In January 1996 WHO's governing board recommended that all stocks of smallpox be destroyed by 1999.

Symptoms and effects

Despite the widespread availability of a vaccine, the potential weaponisation of variola continues to pose a military threat because of the aerosol infectivity of the virus and the development of susceptible populations. The incubation period for smallpox averages 12 days, and exposed persons are quarantined for a minimum of 16 to 17 days following exposure. **Clinical manifestations begin acutely with malaise, fever, rigors, vomiting, headaches and backaches**; 15 per cent of patients develop delirium. Around 2 to 3 days later, an enanthem appears concomitantly with a discrete rash about the face, hands and forearms. The rash then spreads to the trunk over the next week, and lesions progress quickly from macules to papules, eventually to pustular lesions. From 8 to 14 days after onset, the pustules form scabs which leave depressed depigmented scars upon healing. Smallpox must be distinguished from other vesicular exanthems (fluid-filled eruptions), such as chickenpox, erythema multiform with bullae, or allergic contact dermatitis.

5.10.9 Venezuelan Equine Encephalitis (VEE) virus

Venezuelan Equine Encephalitis		
Symptoms	**Incubation period**	**Treatment**
Headache, fever, dizziness, drowsiness or stupor, tremors or convulsions, severe prostration, occasional paralysis, muscular inco-ordination	1 to 5 days	Supportive therapy only, no specific therapy

Venezuelan Equine Encephalomyelitis (VEE) virus is a strain of the encephalitis neurotropic micro-organism that affects primarily equine animals such as horses and mules.

Symptoms and effects
10 per cent of those infected suffer an overt illness if infected via mosquito bite, 75 per cent if infected via the aerosol route. **After an incubation period of 1 to 5 days, onset of illness is extremely sudden, with generalised malaise, spiking fever, rigors, severe headache, photophobia and myalgia in the legs and lumbrosacral area. Nausea, vomiting, cough, sore throat and diarrhoea may follow.** This acute phase

lasts 24 to 72 hours. A prolonged period of aesthenia and lethargy may follow, with full health and activity regained after 1 to 2 weeks. Approximately 4 per cent of patients during natural epidemics develop signs of central nervous system infection, with convulsions, coma and paralysis.

VEE is almost indistinguishable from other viruses that cause human encephalitis such as St. Louis encephalitis, Eastern Equine and Western Equine Encephalomyelitis, Japanese B-Type encephalitis and Russian Far East encephalitis. Natural cases of VEE occur in Central and South America during the summer and early autumn. VEE is very infective, and the incubation period following inhalation of an infectious aerosol ranges from 1 to 5 days.

In the former US and USSR offensive BW programmes, **VEE virus was weaponised in both liquid and dry forms for aerosol dissemination**. This virus has those properties that make it extremely useful as a BW agent, either by a country or by a terrorist group.

5.10.10 Yellow fever

Viral: Yellow fever		
Symptoms	**Incubation period**	**Treatment**
Range from very mild to malignant, sudden onset of chills, fever, prostration, headache, backache, muscular pain, congestion of mucous membranes, nausea, vomiting, jaundice from liver damage, bleeding from stomach and gums, black vomitus	3 to 6 days, rarely longer	Symptomatic therapy, bed rest and fluids needed for even mildest cases. Immunisation considerations before exposure.

The Yellow fever virus belongs to the genus Flavivirus and **is typically transmitted by mosquitoes**. A safe and effective vaccine is available.

Symptoms and effects
Yellow Fever is a bloodstream infection, naturally transmitted from the bites of the female *Aedes aegypti* mosquito. Yellow Fever is generally not lethal and only mildly incapacitating. **Symptoms of Yellow Fever**

include a sudden onset of chills, headache, fever, muscle pain and prostration. Severe cases also show mucous membrane congestion, gastrointestinal complications, and liver damage from jaundice and haemorrhaging of the stomach. Yellow Fever runs a rapid course of infection, incubating for 3 to 6 days and then ending within 2 weeks in either death or full recovery.

CHAPTER 6: BIOLOGICAL TREATMENT

CHAPTER 6: BIOLOGICAL TREATMENT

The material in this Chapter is provided for the use of qualified medical personnel and as background information for first responders. It is provided as a guide only. It does not supersede national or local procedures and practices.

6.1 Bacteria

6.1.1 Anthrax
A vaccine is available and consists of a series of 6 doses over 18 months with yearly boosters. The first vaccine of the series must be given at least 4 weeks before exposure to the disease. This vaccine, while known to protect against anthrax acquired through the skin in an occupational setting, is also believed to be effective against inhaled spores released during a BW attack.

Unvaccinated **exposed adults should receive antibiotics, with ciprofloxacin, 500 mg bd, or doxycycline 100 mg bd** being the treatment of choice. For post-exposure prophylaxis **the administration of antibiotics should be continued for at least 60 days** in those **exposed and they should be immunised** – that is, receive 3 doses of vaccine before antibiotics are discontinued.

Anthrax

	Ciprofloxacin	Doxycycline	Penicillin
Vaccine Prophylaxis	Bioport Corp Vaccine 0.5 ml @ 0, 2, 4 wks, 6, 12, 18 months then annual boosters		
Drug/ Therapeutic	Ciprofloxacin	Doxycycline	Penicillin
Chemotherapeutic Regimen or Supportive Therapy	Adults: 400 mg IV q 8-12 hours for 60 days. Also 3 doses of vaccine 2 weeks apart if unvaccinated. Children: 20-30 mg/kg IV divided into 2 doses per day for 60 days not to exceed 1g/day.	Adults: 100 mg IV @ 12 hours for 60 days plus vaccination. Children: >45kg- same as adults, <45kg- 2.5 mg/kg every 12 hours (max 200mg/ day)	Adults: 2 million units IV q 2 hours for 60 days plus vaccination. Children: IV (PCN G) <12y/o- 50, units every 6 hours; >12y/o 4 million units every 4 hours

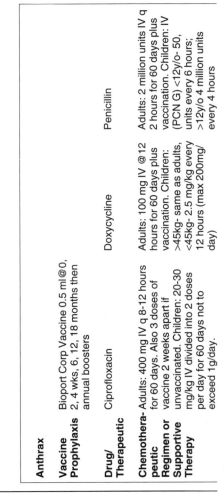

			Children: 25-50 mg/kg divided into 3 to 4 doses per day.
Chemoprophylactic Regimen	Adults: 500 mg p.o. bd for 60 days plus vaccination. Children: 20-30mg/kg divided into 2 doses per day, not to exceed 1 g/day.	Adults: 100 mg p.o. bd × 60 days plus vaccination. Children: >45kg-same as adults; <45kg-2.5 mg/kg every 12 hours (max 200 mg/day)	
			For PCN sensitive organisms
Comments	Ciprofloxacin is the first line of therapy for all patients including pregnant women and children. Alternatives are clindamycin, doxycycline and chloramphenicol. No paediatric doses have been approved for the anthrax vaccine at this time.		

[1] Staining of teeth in foetuses and children <8 with use of tetra or doxycycline has been documented only after >6 cycles of treatment of at least 6 days each.

[2] Ciprofloxacin has been shown to impair cartilage growth in beagles, and therefore is not recommended for use in children or during pregnancy. However, no serious adverse consequences have been reported in children who have used cipro, and it would be recommended for use in a life-threatening situation.

Source: USAMRIID, January 2002.

Unvaccinated exposed **children should also receive antibiotics in the following doses – ciprofloxacin oral or IV dose 20-30 mg/kg divided into 2 doses per day not to exceed 1 g/day; doxycycline oral or IV dose, if greater than 45 kg treat the same as adults, if less than 45 kg give 2.5 mg/kg every 12 hours (max 200 mg/day).**

Cutaneous anthrax can be effectively treated with antibiotics (such as penicillin, tetracycline, ciprofloxacin and doxycycline).

Effective decontamination can be accomplished by boiling contaminated articles in water for 30 minutes or longer, by using some of the common disinfectants. Chlorine is effective in destroying spores and vegetative cells. Again, anthrax spores are stable, able to resist sunlight for several days and able to remain alive in soil and water for years. Consequently, steam under pressure or exposure to dry heat above 284°F (140°C) for 1 hour is necessary to kill all spores.

6.1.2 Brucellosis
Appropriate treatment with antibiotics, particularly by a combination of doxycycline (100 mg orally twice a day in adults and children) and rifampin (600 mg per day in adults and 15-20 mg/kg/day in one or two divided doses in children [maximum 600-900 mg/day]) for at least six weeks. Prolonged combined

therapy is the key. An alternative treatment is streptomycin and the tetracyclines. However, some cases are resistant to all forms of therapy. The relapse rate is high.

Contaminated materials are easily sterilised or disinfected by common methods. For example, pasteurisation is effective for contaminated dairy products.

6.1.3 Cholera
The first consideration in the treatment of cholera is to replenish fluid and electrolyte losses of the body. Intravenous fluid replacement is occasionally needed in patients with persistent vomiting or high rates of stool loss (10mg/kg/h). **Tetracycline therapy (500 mg every 6 hours for 3 days in adults) appears to enhance the effectiveness of rehydration** by reducing the volume and duration of diarrhoea and aiding the disappearance of the organisms.

Other antibiotics for consideration are **ciprofloxacin (500 mg every 12 hours for 3 days in adults)** or **erythromycin (500 mg every 6 hours for 3 days in adults). Doxycycline is recommended for children (6 mg/kg in a single dose [maximum 300 doses])**.

A licensed vaccine is available for use in those considered at risk of exposure. The vaccination schedule is an initial dose followed by another dose 4

Cholera (Toxin)

	Oral Rehydration Therapy	Tetracycline	Doxycycline	Ciprofloxacin	Norfloxacin
Vaccine Prophylaxis	Wyeth-Ayerst Vaccine (50% efficacy, short term) 0.5 ml IM @ 0, 7-30 days, then boosters every 6 months				
Drug/ Therapeutic					
Chemotherapeutic Regimen or Supportive Therapy	During period of high fluid loss	500 mg q 6 hrs × 3 days	Adults: 300 mg once, or 100 mg q 12 hrs × 3 days. Children: 6 mg/kg in a single dose (max 300 doses)	500 mg q 12 hrs × 3 days	400 mg q 12 hrs × 3 days

Chemopro-phylactic Regimen		
Comments	Vaccine not recommended for routine protection in endemic areas. Swedish SBL oral vaccine effective but not available in US	Paediatric treatment includes doxycycline, tetracycline[1], erythromycin, trimethoprim and sulfa, and furazolidone	For tetra/doxy resistant strains

[1] Staining of teeth in foetuses and children < 8 with use of tetra or doxycycline has been documented only after >6 cycles of treatment of at least 6 days each.

Source: USAMRIID, January 2002.

CHAPTER 6: Biological Treatment

weeks later, with booster doses every 6 months. The cholera vaccine currently licensed in the US is no longer given to US military personnel, due to poor efficacy and concerns about complacency. It provides only about 50 per cent protection that lasts for no more than 6 months. There are several new vaccines in testing stages that appear to be much more effective and have a lower side effect profile. However, when these might be available in the US and the UK is not known. **The organism is readily killed by dry heat at 243°F (117°C), by steam, boiling, short exposure to ordinary disinfectants and by chlorination of water**.

6.1.4 Glanders

Mainstream treatments are not yet established and are not as reliable as for other biological agents. Antibiotics (streptomycin and sulphadiazine) have some effectiveness against it. The absence of a glanders vaccine or well-developed spectrum of antibiotics could complicate post-attack treatment efforts even in countries with highly sophisticated medical facilities. In addition, immunity does not follow successful recovery and humans could easily re-contract the disease. **Recommend IV ceftazidime plus trimethoprin/sulphamethoxazole (TMP/SMX) for at least 2 weeks, then switch to an oral therapy (depending on antibiotic susceptibilities) to complete at least a six month course. Regimen for children is ceftazidime at the rate of 50 mg/kg every 8 hours. Because glanders is a difficult disease to**

manage, it is imperative to immediately involve an infectious disease specialist in the patient's management**. While it can withstand dry conditions for up to 2 to 3 weeks, **glanders dies in sunlight within a few hours, as well as from common disinfectants and temperatures over 162°F (72°C).**

6.1.5 Melioidosis

While no known vaccine exists that can counter melioidosis, **a similar treatment regime to glanders is recommended- IV ceftazidime plus trimethoprin/ sulphamehoxazole (TMP/SMX) for at least 2 weeks**, then switch to an oral therapy (depending on antibiotic susceptibilities) to complete at least a six month course. **For children the recommended ceftazidime doses are 50 mg/kg every eight hours**. Because glanders is a difficult disease to manage, **it is imperative to immediately involve an infectious disease specialist in the patient's management. The bacteria are susceptible to moist heat over 165°F (74°C) or chemical disinfectants such as phenol (1 per cent) or formalin (0.5 per cent).**

6.1.6 Plague

Antibiotic choices for post-exposure prophylaxis of pneumonic plague include doxycycline (100 mg orally twice daily for 7 days or the duration of risk of exposure, whichever is longer) and ciprofloxacin with chloramphenicol as an alternative.

Plague

Vaccine Prophylaxis	No vaccine available		
Drug/Therapeutic	Streptomycin	Doxycycline; ciprofloxacin	Chloramphenicol
Chemotherapeutic Regimen or Supportive Therapy	Adults: 30 mg/kg/d IM in divided dose × 10 days Children: 15 mg/kg IM every 12 hours (max dose 2 g daily)	Adults: 200 mg IV then 100 mg q 12 hrs × 10-14 days Children: Cipro- 15mg/kg IV every 12 hours, Doxy- >45 kg same as adult dose; <45 kg 2.2mg/kg every 12 hours (max dose 200 mg/day)	Adults: 1 gm IV q 6 hrs Children: 25 mg/kg 4 times daily.

Chemoprophylactic Regimen	Adults: 100 mg q 12 hrs × 7 or more days Children: Cipro- 15 mg/kg IV every 12 hours, Doxy- >45 kg same as adult dose; <45 kg 2.2mg/kg every 12 hours (max dose 200 mg/day)
Comments	Plague vaccine not protective against aerosol challenge in animal studies
	See anthrax comments for use of doxycycline in pregnant women and children. Alternates: Bactrium or Cipro
	For plague meningitis

Source: USAMRIID, January 2002.

Pneumonic plague can be successfully treated within 24 hours of onset of symptoms. Treatments include **streptomycin (Adults: 30 mg/kg IM, in 2 divided doses daily for 10 days Children: 15 mg/kg IM every 12 hours [max 2 g/day], or doxycycline – Adults: 200 mg IV, then 100 mg IV every 12 hours for 10-14 days, Children: >45 kg same as adults; <45 kg – 2.2 mg/kg IV every 12 hours [max 200 mg/day]) and gentamicin**. Alternative treatments include doxycycline, chloramphenicol and ciprofloxacin.

Decontamination can be effected by boiling, dry heat above 130°F (54°C) steam or treatment with Lysol™ or chloride of lime. The organism is killed by exposure to heat at 130°F (54°C) for 15 minutes or it is killed by exposure to sunlight for 3 to 5 hours.

6.1.7 Tularemia, Rabbit fever or Deer fly fever

Antibiotics such as **streptomycin (Adults: 1 g every 12 hours intramuscular [IM] for 14 days; Children: 15 mg/kg IM every 12 hours [max dose 2 g daily]) is the treatment of choice. Gentamicin is also effective (Adults: 3 to 5 mg/kg/day for 10 to 14 days; Children: 2.5 mg/kg IM or IV every 8 hours)**. Tetracycline and chloramphenicol treatment are effective as well, but are associated with a significant relapse rate. A live, attenuated tularemia vaccine is available as an investigational new drug. It has been administered to more than 5, persons without significant adverse reactions and is of proven effectiveness in preventing laboratory-acquired respiratory tularemia.

Tularemia

Vaccine Prophylaxis	LVS – Live Attenuated Vaccine (IND) – one dose by scarification		
Drug/Therapeutic	Streptomycin	Gentamycin	Doxycycline
Chemotherapeutic Regimen or Supportive Therapy	Adults: 1 gm IM q 12 hrs × 10-14 days Children: 15 mg/kg IM every 12 hours (max dose 2 g daily)	Adults: 3-5 mg/kg/day × 10-14 days Children: 2.5 mg/kg IM or IV every 8 hours	
Chemoprophylactic Regimen			Adults: 100 mg p.o.q 12 hrs × 14 days Children: same as adult dose; >45 kg 2.2 mg/kg every 12 hours (max dose 200 mg/day)
Comments			Can also use tetracycline

Source: USAMRIID, January 2002.

6.1.8 Typhoid Fever

Prompt use of appropriate antibiotics (ciprofloxacin or ceftriaxone) shortens the period of communicability and rapidly cures the disease. Use of antibiotics varies with the susceptibility of the organism. **Effective decontamination measures include pasteurisation, exposure to heat at 132°F (56°C) for 20 minutes, exposure to 5 per cent phenol or 1:500 bichloride of mercury for minutes, cooking and boiling**. Immunity can be achieved with a live-attenuated oral vaccine which is good for 5 years or a polysaccharide vaccine which is good for 2 years. Doses for children are as follows:

◆ Ceftriaxone 50-75 mg/kg once or twice per day (max 2 g/day), until susceptibilities known, is reasonable

◆ Vaccine – must be at least 6 years old to receive Ty21a oral vaccine. Must be at least 2 years old to receive ViCPS polysaccharide IM vaccine

6.2 Rickettsia

6.2.1 Endemic Typhus/Epidemic Typhus

The course of epidemic typhus can be shortened by the use of antibiotics (doxycycline and chloramphenicol). Supportive treatment and prevention of secondary infections are essential.

Both epidemic and endemic typhus can be destroyed by heat at 112°F (44°C) for 15 to 30 minutes, and inactivated by use of 0.1 per cent formalin and 0.5 per cent phenol.

6.2.2 Q Fever

Tetracycline (500 mg every 6 hours) or **doxycycline (Adults: 100 mg every 12 hours for 5 to 7 days, Children: >45 kg same as adult dose; <45 kg 2.2 mg/kg every 12 hours [max dose 200 mg/day] treat until 72 hours afebrile [7-10 days])** is the treatment of choice, although a combination of **erythromycin (500 mg every 6 hours) plus rifampin (600 mg per day)** is also effective. **Treatment with tetracycline during the incubation period will delay but not prevent the onset of illness.** Vaccination with a single dose of a killed suspension of *C.burnetii* provides complete protection (for at least 5 years) against naturally occurring Q-fever and greater than 90 per cent protection against experimental aerosol exposure in human volunteers. Administration of this vaccine in immune individuals, however, may cause severe cutaneous reactions including necrosis at the inoculation site. This is an Investigational New Drug (IND) product and is not readily available. Newer vaccines are under development.

Q Fever

Vaccine Prophylaxis	IND 610-inactivated whole cell vaccine given as single 0.5 ml s.c. injection	Q-Vax (CSL Ltd, Parkville, Victoria, Australia)
Drug/Therapeutic	Tetracycline	Doxycycline
Chemotherapeutic Regimen or Supportive Therapy	500 mg oral q 6 hrs × 5 to 7 days	Adults: 100 mg oral q 12 hrs × 5 to 7 days Children: >45 kg same as adult dose; <45 kg 2.2 mg/kg every 12 hours (max dose 200 mg/day), treat until 72 hours afebrile (7 to 10 days treatment)
Chemoprophylactic Regimen	Useful for prophylaxis. Start 8 to 12 days post-exposure × 5 days	Useful for prophylaxis – same regimen
Comments	Currently testing vaccine to determine the necessity of skin testing before use	Licensed vaccine in Australia

Source: USAMRIID, January 2002.

6.3 Toxins

6.3.1 Botulinum
Treatment of botulism consists of ventilation assistance and the use of the antitoxin. **The antitoxin, if available, (10 ml IV over 20 minutes for both adults and children)** should be administered immediately when botulism is suspected. Upon recognising a case of botulism, immediately search for all other exposed persons. Treatment after severe set-in is usually ineffective; the antitoxin will not reverse existing paralysis. Recovery is very slow and several months may pass before a victim regains certain muscle movements.

A vaccine which is a pentavalent toxoid of *Clostridium botulinum* toxin types A, B, C, D and E is available under IND status, and this product has been administered to several thousand volunteers and occupationally at-risk workers.

Use of a 1 to 2 per cent hypochlorite solution (five parts water, nine parts bleach) can destroy the toxin. Since the toxin is also sensitive to heat, boiling for 15 minutes or, when in food, cooking for 30 minutes, at 176°F (80°C) will destroy it.

6.3.2 Ricin
There is currently no vaccine to be given before exposure to ricin, although the US is conducting

research on developing such a vaccine. There is also no ricin antitoxin to be given after exposure and the only treatment available involves management of the effects based on the method of exposure to the toxin. If ricin is inhaled, oxygen may be given as well as drugs to reduce inflammation and support the function of the heart and circulatory system. Ingestion of ricin would be treated by emptying the stomach by using a charcoal lavage. **Replacing lost fluid is critical, as fluid losses of up to 2.5 litres are probable**. There is no specific treatment for injected ricin other than making the person comfortable and using any measures normally used to treat organ failure.

Use soap and water to remove contamination from personnel, equipment and supplies.

Ricin Toxin

Vaccine Prophylaxis	No vaccine available
Drug/Therapeutic	No specific antitoxin
Chemotherapeutic Regimen or Supportive Therapy	Inhalation: supportive therapy for acute lung injury and pulmonary oedema. G-1: gastric lavage, superactivated charcoal, cathartics
Chemoprophylactic Regimen	
Comments	

Source: Medical Management of Biological Casualties, USAMRIID, August 1996.

6.3.3 Saxitoxin
Induce vomiting and provide general supportive care. Artificial respiration may be necessary. **Caution: misidentification of this toxin as a nerve gas, with the resultant use of atropine, would increase fatalities**.

6.3.4 Staphylococcus Enterotoxin B (SEB)

No vaccine or antitoxin is available to treat SEB before or after exposure. **The treatment for SEB once symptoms appear consists of pain relievers and cough suppressants**. Additional drug therapies are under investigation. For severe cases (which are expected to be rare) more extensive hospital procedures may be needed such as mechanical breathing and replacement of fluid.

Staphylococcus Enterotoxin B

Vaccine Prophylaxis	No vaccine available
Drug/Therapeutic	No specific antitoxin
Chemotherapeutic Regimen or Supportive Therapy	Ventilation support for inhalation exposure
Chemoprophylactic Regimen	
Comments	

Source: Medical Management of Biological Casualties, USAMRIID, August 1996.

6.3.5 Trichothecene Mycotoxins

No specific antidote or therapeutic regimen is currently available, and all therapy is symptomatic and supportive. Also, there are no available vaccines or chemoprotective pretreatments available.

T-2 Mycotoxins

Vaccine Prophylaxis	No vaccine available
Drug/Therapeutic	No specific antitoxin
Chemotherapeutic Regimen or Supportive Therapy	
Chemoprophylactic Regimen	Decontamination of clothing and skin
Comments	

Source: Medical Management of Biological Casualties, USAMRIID, August 1996.

6.4 Viruses

6.4.1 Dengue Fever

At present, there is no specific treatment for dengue fever and no current vaccines exist to combat the virus. **The only method of controlling or preventing dengue is to combat the vector mosquito, the *Aedes aegypti*.**

6.4.2 The Marburg

No vaccines exist that can counter the effects of this haemorrhagic fever. Supportive treatment is strongly

Marburg/Ebola Virus	
Vaccine Prophylaxis	No vaccine available
Drug/Therapeutic	No specific anti-viral
Chemotherapeutic Regimen or Supportive Therapy	Supportive therapy; shock, hypotension and DIC management
Chemoprophylactic Regimen	
Comments	Isolation and barrier nursing required

Source: Medical Management of Biological Casualties, USAMRIID, August 1996.

recommended. **Suspected cases should be isolated from other patients and strict barrier nursing techniques should be practised**.

6.4.3 Rift Valley Fever

RVF is sensitive to ribavirin *in vitro* and in rodent models. No studies have been performed in human or the more realistic monkey model to ascertain whether administration to an acutely ill patient would be of benefit. **It would be reasonable to treat patients with early signs of haemorrhagic fever with intravenous ribavirin** (30 mg/kg IV initial dose, then 15 mg/kg IV q 6 hours × 4 days, then 7.5 mg/kg q 8 hours × 6 days). This regimen is safe and effective in haemorrhagic fevers caused by some viruses, although a reversible anaemia may appear. Therapy may be stopped 2 to 3 days after improvement begins or antibodies appear.

Avoidance of mosquitoes and contact with fresh blood from domestic animals, and respiratory protection from small particle aerosols are the mainstays of prevention. An effective inactivated vaccine (an IND product) is available in limited quantities. The first dose, 1 ml, would be followed by 1 ml doses on days 7 and 28; exact timing is not critical. Protective antibodies begin to appear within 10 to 14 days and last for a year, at which time a 1 ml booster should be given. A single injection is probably not protective, but two inoculations may provide marginal short-term protection.

6.4.4 Smallpox

Smallpox vaccine is only available through IND (investigational new drug) inoculation with a bifurcated needle, a process that has become known as

---

Smallpox Virus

Vaccine Prophylaxis	Wyeth vaccine – 1 dose by scarification – booster q 3 years for V. major
Drug/Therapeutic	Vaccinia Immune Globulin (VIG)
Chemotherapeutic Regimen or Supportive Therapy	
Chemoprophylactic Regimen	0.6 ml/kg IM (within 3 days of exposure, best within 24 hours)
Comments	Pre- and post-exposure vaccination recommended if 3 years since last vaccine.

Source: USAMRIID, January 2002.

---

scarification because of the permanent scar that results. It is effective if given up to 1 week post-exposure. **Protection from the vaccine lasts 3 to 10 years. After this length of time it is recommended that booster doses be given if exposure has occurred**. Vaccinia immune globulin can be given to those who cannot receive the vaccine because of contraindications (pregnancy, immunosuppression). The dose is 0.6 ml/kg and administration immediately after or within the first 24 hours of exposure would provide the highest level of protection.

6.4.5 Venezuelan Equine Encephalitis (VEE)

Only supportive treatments exist as the best aids to recovery because antibiotics and chemotherapy are not effective against VEE. Small-scale vaccinations have been employed, but widespread vaccines have not yet been developed. A vaccine (under IND use), designated TC-83, is a live, attenuated cell-culture-propagated vaccine which has been used in several thousand persons to prevent laboratory infections. The vaccine is given as a single 0.5 ml subcutaneous dose and is not useful for post-exposure prophylaxis.

Venezuelan Equine Encephalitis

Vaccine Prophylaxis	TC-83 live attenuated vaccine (IND) 0.5 ml s.c. × 1 dose (vaccine not approved for children)	C-84 (formalin inactivation of TC-83 [IND]) 0.5 ml s.c. for up to 3 doses q two weeks (vaccine not approved for children)
Drug/ Therapeutic	No specific anti-viral	
Chemothera-peutic Regimen or Supportive Therapy	Supportive therapy, analgesics, anticonvulsants	
Chemopro-phylactic Regimen		
Comments	Reactogenic in 20%. No serocon version in 20%	Vaccine used for non-responders for TC-83

Source: USAMRIID, January 2002.

6.4.6 Yellow Fever

General supportive treatment such as rest and fluids are effective. Recovery provides lasting immunity, and there are no known second attacks. During epidemics, case fatality rates may exceed 50 per cent, although usually they are less than 5 per cent. **A licensed vaccine is very effective; boosters are required every 10 years. In children, this vaccine is only recommended for those at risk and older than 9 months of age but can be given to those as young as 4 months.**

6.5 Management of biological casualties

One of the first and most important requirements faced by the medical community is to **determine if the initial casualties are caused by a natural outbreak of disease or whether the casualties are from a biological warfare attack**. The scale of response required in the event of a successful bio-attack is likely to involve a range of health care providers. These may include:

◆ Doctors
◆ Paramedics
◆ Veterinary surgeons
◆ Nurses
◆ Dentists
◆ Microbiologists
◆ Public health officials
◆ Officials from the relevant government ministries
◆ Military personnel

Self-protection is the first priority upon arrival at a scene where biological agents are suspected to have been disseminated. The rescuer or the medical attendant must have adequate protection before entering a contaminated area. This may involve the use of a mask or respirator and could involve prophylactic treatment with antibiotics or other drugs. In addition to the mask or respirator, appropriate gloves and overgarments would be necessary to protect the individual, at least in the early stages, until the biological agent could be identified.

Casualties from biological warfare will require the same type of treatment as those who have acquired the disease by natural means. The methods of providing treatment will differ significantly from those of a natural outbreak, but the basic principles will be the same. **The biggest problem will be providing medical care to such a large number of cases within a short period of time**.

Vulnerability of the urban environment
The following 'mathematical modelling' of tularemia illustrates the problem of providing treatment to mass casualties. "An intermediate-sized strategic hypothetical target could be represented by the typical city and its surrounding network of suburbs – about 400 to 800 sq miles. One jet fighter could disseminate 100 gallons of *Francisella tularensis* (the causative bacterium of tularemia) along a 50 mile line.

Modelling with a US example suggested that such an attack would be expected to produce 50 per cent casualties in 4 to 7 days in metropolitan Washington, DC. It was also estimated that severe pneumonia would afflict 1.9 million people, of whom 20 per cent would be expected to die (without antibiotic treatment). This level of casualties would destroy the ability of a significant and vital section of the US and would severely stress the remainder of the country to provide the resources needed to limit the biological disaster." [From: JAVMA, Vol. 190, No. 6, 1987, p. 716].

Conversely, a smaller-scale covert biological attack could be directed towards a key administrative centre, institution or installation, whether military or civilian. In this scenario, far fewer casualties might be expected. However, a successful attack could produce sufficient casualties to overwhelm local medical facilities.

The problems of initial clinical diagnosis in a mass attack would be great, particularly since the mode of entry (aerosol route) and dose of the infectious agent may be principally different from a natural epidemic situation. A further problem, adding to the diagnostic difficulties, is the fact that the **agent or the disease would often be unfamiliar to the medical profession in a particular geographic area**.

Medical Care

It is anticipated that the vast majority of patients from a biological warfare strike will not require special equipment such as elaborate x-ray facilities, oxygen tents and specialised surgical services. An important exception is a situation resulting from the use of botulinum and saxitoxin where dramatic acute symptoms such as respiratory paralysis would indicate the need for various types of advanced equipment.

The type of response information required by a targeted population depends on the agent.

If the disseminated organism produces an illness that results in relatively few deaths (Venezuelan Equine Encephalomyelitis), home or group medical care will be quite effective. A disease for which specific therapy such as antibiotics is indicated (Q fever or tularemia), would need instructions relative to obtaining and administering the drug. If a disease with a high lethality rate and for which there is no specific therapy is encountered, instruction for general supportive care that might be provided by non-medical personnel could be disseminated. This would not be a very satisfactory effort, but it would be the best available under the circumstances. There simply is no satisfactory answer to the problems of management of casualties in this latter situation.

The response of the infection to antibiotics would be the same regardless of how the patient contracted the disease. The enemy might use an antibiotic-resistant strain of a particular micro-organism, but the source or route of the infection will not change the sensitivity of the organism or the response of the patient.

Gradual onset of illness

Individuals becoming ill from an attack with a biological weapon would not all become casualties at the same time as they would, for example, in the case of a nuclear explosion, saturation bombing or massive surprise attack with nerve gas. **Dosage variation and host resistance cause the onset of illness to be spread over a number of hours or days**. Thus, the casualty load would occur as represented roughly by a bell-shaped curve with relatively few casualties at the beginning, the number becoming higher over successive hours or days until a peak is reached. An exception to this aspect might be an attack with rapidly acting biological toxins, for example, SEB.

Those infected by a biological agent would retain their physical abilities for a period of time after the attack (incubation period) and thus theoretically could return home without assistance or elaborate evacuation systems such as needed in other mass casualty situations. However, such a return might not be advisable for epidemiological reasons (for example, person-to-person transmission with smallpox or pneumonic plague), at least not until an aetiological diagnosis had been established. Thus, under certain conditions, the situation might have to be "frozen" by the isolation of groups of personnel in their location together with one or more medical personnel.

It may be necessary for one physician, with a small number of ancillary medical personnel, to care for several hundred patients. Information could be disseminated about the normal course of the disease, specific signs or symptoms of adverse prognostic significance to watch for, when to seek individual medical attention or advice and how to obtain essential medical supplies. This would allow a limited number of professional personnel to care for the maximum number of patients. A particular problem would be created by the appearance of cases of secondary infection. Such cases might, on the other hand, assist in the clinical and laboratory-based diagnosis.

Decontamination activities of medical management

Although most activities of medical management would take place a considerable time after the biological attack itself, physical measures such as decontamination and collective protection will still have to be considered. Beyond the general principles of decontamination and disinfection, certain practical applications should be considered. It is highly questionable whether large-scale decontamination of material, equipment and buildings is a realistic possibility. If it should be judged necessary, however, the best methods in most instances would be disinfection with hypochlorite, paraformaldehyde or ethylene oxide and liquid disinfectant applied with specially designed equipment. Special precautions would need to be taken with the use of these disinfectants, because they are highly toxic to human beings and animals and can damage electronic components.

Given the diversity of potential biological agents, **the single most important action in decontamination of exposed personnel would be a soap and water shower**. Clothing should be disinfected by common methods such as boiling and soaking in chlorine solution.

To protect against secondary transmission of diseases, there also should be adequate stores of pesticides and insecticides for control of insects and vermin that might contribute to secondary spread.

Public order and public reassurance
An essential aspect of medical management following a BW attack would be to allay panic and support the affected community. This could be achieved most effectively if the public could be assured that:

◆ The cause of the illness is known
◆ The course of the disease can be described with reasonable accuracy
◆ The outcome can be predicted.

This type of assurance could be provided only if an accurate aetiologic diagnosis can be made shortly after the onset of illness. If this assurance cannot be provided, the reaction of the public might create greater problems than the disease itself. Such rapid diagnosis would comprise two phases, the first being based on clinical observation and epidemiological indications only, the second being based on precise laboratory findings. It should be remembered that special difficulties may arise. Recognition may be complicated by an exotic agent that is difficult for medical personnel to identify. A genetically modified agent would pose similar difficulties in diagnosis.

There is little question that many deaths would occur if an effective attack with a biological agent were mounted. This would be true even if the so-called incapacitating agents were employed. In this case, deaths would occur particularly among the very young, the very old, and those suffering from underlying disease or some other form of weakening stress. As deadly as a BW attack might be, the panic associated with the attack, once it becomes known, may be much worse than the damage from the actual attack. **The real 'force multiplier' in BW is the panic, misinformation and paranoia associated with it**.

Approach to a BW munition

Once a possible BW device is located, first responders who are not qualified EOD operators should restrict their actions to:

(1) Establishing a zone of exclusion
(2) Notification of the correct authorities (using the appropriate mnemonic where appropriate)
(3) Population protection measures in case of release.

Munitions (point-source bomblets, line source tanks or crude garden sprayer-type systems, for example) **should only be approached by qualified EOD personnel.** Moreover, the device could also contain an explosive or gaseous penalty.

All members of the responding team should be wearing the appropriate level of PPE.

CHAPTER 7: POST INCIDENT

CHAPTER 7: POST INCIDENT

Experience drawn from acts of *conventional* terrorism, especially attacks involving large bombs in vehicles, have highlighted the impact mass terrorism can have on communities. A chemical or biological attack, because of uncertainties associated with the long-term consequences and the potential magnitude of the initial shock, is likely to have even greater impact. However, despite obvious differences it must be appreciated that, as in a conventional scenario, the incident location will remain a crime scene until the on-site investigation has been completed.

It is important to understand that the way in which responders act on the scene and afterwards can do much to minimise the overall psychological effect of a catastrophic incident. It is also important to recognise the symptoms of severe trauma in order to differentiate them from the likely symptoms of a chemical or biological exposure. In terms of post-incident considerations, the following factors are pertinent:

◆ Incident scenes are also crime scenes and materials within the event area may be considered as evidence and therefore subject to a number of special handling restrictions.
◆ Much can be learned operationally from the management of a chemical or biological event, as such a comprehensive after-action review of the

situation, with follow-on reports and documentation, will assist others in planning for future events.
◆ Media and public relations in the wake of catastrophe, while not necessarily a priority for first responders, must also be considered.

7.1 Psychological impact

'Communities'

Often communities may respond to a catastrophic event by undergoing the following phases:
◆ **Initial Impact (*response stage*)**: Shock and disbelief tend to be the first reactions to a catastrophic event. People may feel a sense of dislocation as the social order is severely disrupted by injury and loss of life.
◆ **Heroic Period**: Response personnel acting on the scene represent the heroic period of a catastrophe. Responders provide psychological support for the victims by their mere presence and through their acts of care and support for victims. Altruistic acts by first responders and the communities who mobilise to assist the victims provide the necessary support during catastrophe.
◆ **Honeymoon Period (*consolidation phase*)**: For a brief time, rivalries and conflicts are forgotten as the wider community comes together to face the crisis. Radical change is possible during this time as the disaster provides the impetus for a community to adapt.

◆ **Disillusionment**: When attention turns away from the crisis and the initial intervention period ends, the survivors and residents are left to deal with the impact of trauma and loss. Old conflicts return and dealing with the terror of the catastrophe may seem unmanageable as depression sets in.

◆ **Reconstruction (*recovery and restoration of normality*)**: After a catastrophic event, new steps must be taken to show a symbolic departure from the incident. The community cannot heal psychologically from an attack if a 'business as usual' attitude ensues. Effective and visible leadership is key during this phase, as leaders must focus on actions that make progress toward preventing future tragedy and overcoming the current loss.

Individual psychological impact

Violence and trauma may cause a range of physical, emotional and behavioural responses. Such reactions may have immediate onset or occur over a longer period. Supervisory officers should be aware of these possible changes, not just in relation to their subordinates, but within a wider frame of reference including friends, colleagues and (not least of all) self awareness. In relation to conventional acts of terror, some research suggests that approximately one out of every five 'rescue personnel' required dedicated aftercare provision.

Physical symptoms
◆ Agitation and hyperarousal
◆ Heart palpitations
◆ High blood pressure
◆ Sweating
◆ Adrenaline rush
◆ Gastrointestinal distress
◆ Recurrence of allergies, asthma or eczema
◆ Hot or cold flushes
◆ Loss of sleep or appetite
◆ Tightness in the throat or chest
◆ Fatigue

Behavioural changes
◆ Insomnia or nightmares
◆ Hypervigilance
◆ Easily startled
◆ Isolation, withdrawal
◆ Inability to experience pleasure in daily activities

Emotional changes
◆ Anxiety and generalised fear
◆ Increased self-doubt
◆ Irritability
◆ Anger or rage
◆ Sadness
◆ Grief or depression
◆ Numbness or blunted effect
◆ Hopelessness and helplessness
◆ Despair and defeat
◆ Survivor guilt

Cognitive changes
◆ Decreased ability to cope with daily life
◆ Difficulty making decisions
◆ Memory loss
◆ Frequent confusion
◆ Decreased ability to take in new information
◆ Decreased ability to do computations
◆ Intrusive thoughts and images of the violence or its aftermath
◆ Questioning of faith or loss of spiritual beliefs

Intervention and support
The role of occupational health advisers and other health professionals is likely to be paramount in determining the appropriate intervention strategy. In addition to formal programmes, such as Critical Incident Stress Debrief (CISD) sessions, the following considerations may prove useful as part of any local initiative:
◆ Verbal reassurance and focus on coping behaviours that re-establish physical and emotional safety
◆ Identification of constructive coping behaviours and restorative activities
◆ Discourage coping through use or abuse of substances
◆ Reinforcement of positive health habits and regular exercise

◆ Emphasise quiet rest, comfort, food and the company of good friends

◆ Maintain a calm routine at home and at work.

Recovery from a traumatic incident may take months and years depending on the individual(s) concerned and the type of incident. Recovery may be further complicated by a variety of factors including events triggered by subsequent traumatic events.

7.2 The after action review

After response operations are completed, it is important that responders have the opportunity to provide feedback. Typically, this has been found most effective in groups of between four and ten. While any debrief/feedback sessions are likely to be co-ordinated by occupational health advisers or other appropriately qualified bodies, local discussion periods may include:

(a) **What was done right** – this is an opportunity to emphasise the successes of the operation and to give praise where praise is due. It may also highlight which practices were put in place that were most effective and indicate how successes can be built on during training.

(b) **What was done wrong** – note simply a list of mistakes, but have a discussion concerning accidents, near misses and collective failures.

(c) **How effectiveness can be improved** – emphasising processes that can be implemented to overcome mistakes.

7.3 Public relations/media considerations

Responders are unlikely to become directly involved in preparing a media strategy. However, it should be recognised that the collation and distribution of information after a chemical and biological incident is critical in ensuring a rapid return to normality. The public has a need to know what occurred. Lack of information can be just as panic inducing as information that is *exaggerated*.

Responders are therefore likely to be asked to supply Silver or Gold commanders with accurate and timely information from the scene and should be prepared to meet this requirement. It is equally important that information is provided 'from the centre' and not from the inadvertent or injudicious comments of first responders, made to the media directly.

To provide press and other briefings, Silver or Gold commanders will require the following information:
◆ **Accurate details** – information that is inaccurate heightens uncertainty and fear.
◆ **Regular updates** – especially if circumstances change. Information provided at the right time can serve to minimise hysteria, rumour and misinformation.

It is vital that response personnel should be given instructions on how to conduct themselves if contacted by reporters. If brought on side, the media can be the best way to inform the public of actions to take during a crisis and post-incident.

7.4 Post-incident checklist (supervisor's checklist for post-incident care)

◆ **Self-assessment**: As a supervisor, you must ensure you are mentally and physically able to deal with post-incident events

◆ **Post-incident review/critique:** Conduct an after-action review/critique of the technical aspects of the incident. Allow all first responders to express their successes and bring forward issues that need to be addressed. Assign personnel to follow up on concerns and ensure timely action or updates.

◆ **Post-incident recovery – Questions:**
- What do I need to do?
- What do I need to follow up ?
- How are the responders for whom I am responsible doing?
- Who responded?
- How did we respond?
- Actions in route?
- Upon arrival on-scene?
- During incident?
- Were other agencies involved in response?
- Actions taken?
- Mitigation techniques employed?
- Substance involved/type of event, chemical, biological or hazardous material?

- Recovery operations? Who was responsible? Actions taken?
- Follow-up actions required?
- Post-incident recovery? What do I need to do, what do I need to follow up on? How are my responders coping?

◆ **Observe:** Watch assigned responders for signs of stress and task saturation which could induce stress during the post-incident time frame

◆ **Personal Touch:** Hold informal meetings with assigned personnel. Ensure those with command/supervisory responsibility, or other key personnel, are talking with their assigned personnel. Ensure your entire team is looking actively for signs of post-incident stress.

◆ **'Buddy' System:** For a specific period it may be beneficial to allow assigned personnel to work and conduct recovery operations in what is sometimes known as a 'buddy' team. This will allow those who are feeling stress to share their thoughts and enable those who are exhibiting stress to be more readily identified.

◆ **Counselling:** Involve occupational health advisers at the earliest opportunity after an incident and conduct periodic follow-up counselling sessions as needed. (Signs of stress may only become apparent some time after the event.)

◆ **Medical:** In consultation with occupational health advisers or other appropriate health professionals, arrange medicals for all personnel involved in the incident. This will help to ensure that there are no undiscovered side affects related to the response to an incident. The benefits of this action can be summarised thus: first responders will be reassured that they are not further exposing their family/community. This will ease the minds of the responders and their families.

◆ **Equipment:** More often than not, equipment recovery is overlooked due to pressing issues at hand
  • Conduct equipment inspections
  • Ensure equipment has been properly cleaned/decontaminated
  • If necessary, have equipment recovered or replaced as soon as possible
  • Following any After-Action Review (AAR), see what equipment was effective and what equipment needs to be upgraded. (This form of activity is vital to ensure the confidence of responders is maintained.)

7.5 Post-incident checklist (supervisor's checklist for community post-incident response)

The continued support of affected or vulnerable communities is another essential component in ensuring a rapid return to normality. Again, community liaison is not a unilateral concern of emergency or first responders. However, because of the close community links that are already likely to be in place, first responders should be aware of the following issues.

◆ Community presence – maintaining high levels of visibility will serve to reassure the wider community

◆ Responders may be called upon to participate in public affairs programmes

◆ Pass along as much relevant information as possible through the appropriate chain of command as soon as possible

◆ In accordance with the community liaison plan, ensure accurate and timely information is distributed to local community leaders as well as members of your response unit

◆ Control rumours by addressing them head on and in a timely manner. (Do not let them take on a life of their own. Ensure serious issues are passed along the chain of command rapidly.)

◆ Work with the community to ensure a timely and co-ordinated recovery. Be prepared to become involved; the community will look to you for advice.

◆ Ensure all appropriate agencies are being, or have been, notified about the affects of the incident

◆ It may be necessary to 'follow-up' with agencies involved in the incident to ensure proper mediation and recovery operations have been put in place. For example:

  • Water supply utilities
    – Did they clearly understand issues concerning run-off?
    – Did they receive a list of all chemicals and their properties in order to purge public water systems?

- – Have they posted a restriction on the water supply and a time frame?
- Hospitals and other emergency health care providers
  - – Were the local hospitals notified of the potential effects of the substances involved?
  - – Were medical personnel screened for residual affects?
  - – Have initial receiving points at medical facilities been properly decontaminated and checked by qualified personnel for contamination?
  - – Were decontaminated clothing and personal effects transferred to the appropriate agency for crime scene evidence or disposal?
- Police and rescue personnel
  - – Has the medical and psychological screening of response personnel been initiated? (What is the take-up rate?)
  - – Ensure personnel are briefed on possible long-term exposure affects.
  - – Report and/or resolve interagency conflicts quickly.

CHAPTER 8: APPENDIX

CHAPTER 8: APPENDICES

8.1 (A) Detection equipment

Specialist equipment available to first responders is likely to be limited to those capable of detecting chemical materials only. The process is commonly achieved by *monitoring* the air for the presence of specific vapours, pre-set levels of which trigger an alarm. Within the military, and increasingly within the first responder community, the detection, identification and monitoring of chemical vapour hazards is undertaken using a variety of hand-held systems. In terms of identifying liquid contamination (in a battlefield scenario this might be in the form of droplets following a hostile over-flight or bombardment), the most common means is by detector 'paper'. In the context of a first responder scenario, the problem with detector paper is that it is most likely to show that the user is already contaminated – rather than what the user needs to avoid. In relation to unopened containers, portable x-ray systems (especially those using 'false colour') and a system under development by SAIC, Pulsed Elemental Analysis using Neutrons (PELAN), can discern liquid material within a sealed container. PELAN can even distinguish between explosives and CW agents.

Biological detection

The detection of biological materials is problematic and is usually considered reliable only when conducted in a laboratory by a specialist agency. A system called

RAPID (Ruggedised Advanced Pathogen Identification Device) produced by Idaho Technology Inc is one of the few 'portable' systems available currently. It utilises thermal cycling and real-time fluorometer technology. However, its application in a first responder scenario has yet to be proven.

In relation to 'white powder' incidents, where a quantity of suspected biological material is associated with a threat, relatively inexpensive field assay equipment may be of some use. Typically, this technique involves adding the suspicious material to a reagent and is similar in some respects to a 'home pregnancy test kit'. Commercially available systems (the Bio-Threat-Alert test kit is produced by Tetracore, for example) claim to react reliably to the presence of a variety of biological materials, including B. *anthracis* (the causative agent of anthrax). On this basis, the technology should be regarded as a means of indicating the likely status of a physical entity, rather than a detector.

Detector papers (colour change chemistry)

Detector papers are used typically to identify the presence of a chemical contact hazard. They may be attached to PPE (usually on the extremities; footwear, helmets, knees, elbows) and items of equipment – including prodding implements. Papers provide a visual indication by changing colour when in contact with certain liquid agents (typically nerve, blister and blood).

Monitors

A wide range of chemical agent monitors is available commercially. In relation to equipment available to first responders, the terms *monitor* and *detector* are often used interchangeably. However, it must be remembered that however such equipment is described, it is doing no more than sampling air collected at its nozzle – which is typically at the end of a first responder's arm. The equipment is actually providing no more than an indication of what materials are being sampled, not what materials may be nearby, but have yet to be sampled. The term detector must therefore be used with caution.

User considerations:

◆ Can the system detect a range of chemical materials?

◆ Can it detect several materials simultaneously – or does the operator have to change detection modes manually?

◆ Can it detect poisonous materials at pre-lethal concentrations? (that is, could it be used effectively as an *early warning* system for unprotected officers?)

◆ Does it give an indication of concentration? Ideally, this will be in mg/m^3

◆ How likely is it to produce false positives or false negatives?*

◆ How does the equipment cope with interferents? (smoke, dust, cosmetics and so on)

◆ What is the battery life?
◆ Does it take readily available batteries?
◆ Does it require a carrier gas?
◆ Does it contain a radioactive source? (Will a special licence be required?)
◆ What is the maintenance burden?
◆ Is the equipment affected adversely by extended periods in storage?
◆ What is its operational life expectancy?
◆ Is it compatible with the equipment used by other agencies?
◆ Can it be used in low/no light conditions

(* false positive = an alarm without a threat; false negative = the failure to detect a threat. False positives are irritating and adversely affect operator confidence in the equipment. False negatives may have lethal consequences.)

Equipment available to first responders may be classified under one of three headings:
◆ **Personal monitors**
◆ **Team monitors**
◆ **Forensic analysis**

The detection technologies vary from system to system and include:
◆ Ion Mobility Spectrometry (IMS)
◆ Surface Acoustic Wave (SAW)
◆ Gas Chromatography (GC)

◆ Mass Spectrometry (MS)
◆ Flame Spectrophotometer (FS)

In theory, by combining two or more detection technologies at the scene of a suspected release, it should be possible to discount false positives with greater confidence.

Personal monitors

Personal monitors are lightweight monitoring systems, suitable for attachment to a responder's clothing/ equipment. These systems are relatively new and have yet to be proven in an operational environment. However, test results provided by some manufacturers, especially those following independent testing, are encouraging.

Use: the equipment is most suitable for use by officers whose role **may** bring them into contact with a vapour hazard, but whose function is not primarily one of detection. This may include:
◆ Firearms officers
◆ Search team members
◆ Paramedics

Equipment of this type includes: the BAE Systems JCAD (utilising SAW technology); the SAIC SmallCAD (utilising SAW and IMS technologies) and the Smiths Detection LCAD (utilising IMS technology).

Team monitors

Much of the hand-held equipment used to detect chemical materials was designed against a military specification, primarily to counter the threat of exposure to the most common chemical weapons in the arsenal of the Former Soviet Union (FSU), namely nerve agents and blister agents. However, equipment developed or designed recently is usually able to detect additional materials (typically, blood agents and choking agents). Increasingly, some systems can be programmed by the manufacturer to alarm in the presence of other materials of concern (Toxic Industrial Chemicals and Toxic Industrial Materials).

Use: hand-held systems are most probably of use to first responders whose primary function is the detection of chemical materials. This group may include:

◆ Search team officers
◆ Cordon sentries
◆ Medical personnel responsible for triage
◆ Decontamination teams

Equipment of this type includes: the BAE Systems JCAD (utilising SAW technology); the Bruker IMS 2000 (utilising IMS technology); the Environmental Technologies APD2000 (utilising IMS tehnology); the Giat Industries AP2C (utilising FS technology); the SAIC SmallCAD (utilising SAW and IMS technologies) and the Smiths Detection LCAD and CAM (utilising IMS technology).

Forensic analysis

Detailed analysis of a chemical material is most reliably conducted within a laboratory by a specialist agency. However, some relatively portable systems are produced commercially. In terms of first responder use, it is difficult to see any particular advantage over *Team Monitors*. However, such equipment may be utilised to:

◆ Support Scene Of Crime Officers (SOCOs)
◆ Discount false positives
◆ Support medical specialists in determining the most appropriate casualty treatment regime

An example of a sophisticated system (utilising GC technology) is the Inficon HAPSITE. A less complex system (utilising colour change chemistry) is the Drager 'hand-pump' system.

8.2 (B) Personal protective equipment

Total individual protection requires an integrated approach. The primary protective mechanism is respiratory protection for both chemical and biological agents. A properly fitted protective mask when combined with an overgarment, gloves and boots can provide excellent protection. However, full protection can degrade individual and unit performance by as much as 50 per cent. (Work periods will become progressively shorter; rest periods will become progressively longer.)

Respiratory protection

Respiratory protection devices can be divided into two types:

◆ Air-purifying respirators in which the atmospheric air is purified or filtered before reaching the user

◆ Self-Contained Breathing Apparatus (SCBA), that supplies clean air to the user independent of the surrounding atmosphere. (This may be from bottled supplies or in the form of a *re-breather*).

Air-purifying respirators: the military and most police first responders use negative pressure, full-facepiece air-purifying respirators. Changeable filters remove particulate and gaseous contaminants during inspiration. (The most common design of respirator also contains a drinking tube system.) Respirators must be fitted by competent individuals and maintained to set standards. Responders with respirators should always have immediate access to spare canisters and be practised in canister changing drills.

The Powered Air-Purifying Respirator (PAPR) uses a power source to draw atmospheric air across a gas-particulate canister into the facepiece. This creates a positive pressure inside the facepiece, which provides some protection against leaks in the facepiece. The power source, usually a battery pack, is worn on the belt, and a hose goes to the facepiece itself.

Neither type of respirator can be used in oxygen-deficient environments and both are intended for use in lower concentrations of known contaminants.

Respirators are not always appropriate for use by first responders operating within areas of suspected contamination. Special consideration should be given to the suitability of Level C protection in the following circumstances:

◆ When the nature of the hazard has not been quantified (agent type, concentration and so on)
◆ When the air is oxygen deficient

Self-Contained Breathing Apparatus (SCBA): SCBA provides complete respiratory protection as breathing air enters the facepiece from an independent source, usually an air tank worn on the back. This apparatus has two disadvantages: the weight of the air pack (about 40 lb (18 kg)) is cumbersome, and the air supply is limited (typically, 30 minutes or less). There are specific mandatory actions associated with the maintenance and use of SCBA. To some extent, the limitations of carrying air in tanks can be overcome through the use of rebreathers. These devices recycle the wearers exhaled air through a 'scrubber' and add oxygen from a separate cylinder. Some rebreathers allow a working duration of up to 4 hours.

Dermal Protection
Dermal protective ensembles are used in combination with respirators to protect first responders from vapour and liquid hazards. This is achieved in one of two ways.

◆ First, by wicking liquid contamination across the material and using an inner barrier to 'trap' poisonous materials before they reach the skin. An example of this approach is the Mk4/5 NBC suit.
◆ Second, by using a barrier that toxic material cannot permeate, such as the *Tyvek* suit manufactured by Remploy and a variety of other suppliers

Levels of Protection
Generally accepted designations for levels of protection (excluding military applications) are listed below.

Level A: This ensemble consists of a SCBA or supplied-air respirator with escape cylinder in combination with a fully encapsulating chemical protective suit capable of maintaining a positive air pressure within the suit. This includes both outer and inner chemical-resistant gloves, chemical-resistant steel-toed boots, and two-way radio communications. This ensemble (the "moon suit") is required for the highest level of protection for skin, eyes, and the respiratory system.

Level B: This has the same respiratory protection as level A with hooded chemical-resistant clothing, outer and inner chemical-resistant gloves, chemical-resistant steel-toed boots, and other, optional, items. This does not include a positive-pressure suit. This ensemble is used when the type and atmospheric concentrations of substances have been identified, and these require a high level of respiratory protection but a lesser level of skin protection.

Level C: This is similar to level B except that a full- or half-facepiece air-purifying respirator is worn instead of the SCBA or 'supplied-air' respirator. This should be used when:

◆ The contaminants have been identified
◆ The concentrations measured
◆ An air-purifying respirator is appropriate to remove the contaminants.

Level D: This is a work uniform and provides minimal protection.

Other considerations

In addition to the PPE's ability to withstand certain exposures to C or B materials, consideration must also be given to physiological and psychological factors.

◆ First responders will be unable to wear full PPE for extended periods because of the risks associated with heat exhaustion. This is especially the case when:
  • The weather is hot and/or humid
  • Responders are working in an enclosed environment

- Undertaking relatively strenuous work
- Wearing body armour or carrying equipment
- Breathing through a filter cartridge requires additional effort
- Some responders may be affected by claustrophobia
- Hearing is likely to be impaired because of (a) the wearing of a hood and (b) breathing noises accentuated by the wearing of the respirator/facepiece
- Respirators with eye-pieces restrict the wearer's field of vision
- Special arrangements, drills and procedures will be necessary to minimise risk when taking in fluids or using the lavatory
- Communication with other team members is significantly more difficult

8.3 (C) Zones of operation

Procedures adapted from military protocols describe actions at the scene of a C or B incident in terms of three *Zones* of operation (*hot, warm* and *cold*). In a battlefield environment, each would be UPWIND and UPHILL from the previous zone. In an urban environment, around an office block, shopping centre, airport or railway station for example, it will not always be possible to be upwind and uphill. The priority for responders should be to ensure that large groups of people do not come into contact with contamination (primarily, the *downwind* hazard and any *contact hazard*). Every effort should therefore be made to keep upwind of the suspected scene.

Zones (Hot, Warm and Cold) and cordons –

The terms HOT, WARM and COLD zones are not immediately interchangeable with the more general terms INNER and OUTER cordon. However, at any major incident it is likely that the HOT and WARM zone will be entirely within the inner cordon (see the diagram in Chapter 2). The COLD zone is therefore likely to be located in the area between the inner and outer cordons. The purpose of this appendix is to set out possible actions and activities to be carried out within each zone and during the movement between zones. It is a generic guide and does not supersede any local, national or event specific arrangements.

HOT ZONE

The **Hot Zone is the innermost zone and immediately surrounds the incident site**. It is located within the inner cordon and is where contamination is must likely to be encountered. **Only rescue personnel and EOD personnel with appropriate PPE should enter the Hot Zone.**

Activity in the Hot Zone may involve the following:

◆ **Victims identified, given basic life-saving measures and then transferred to the Warm**

Zone for decontamination. (Only basic life-saving medical care is likely to be given in the Hot Zone. This may be limited to hasty airway control, controlling haemorrhage, and the possible use of antidotes, if available).

◆ **Entry/exit from the Hot Zone should be controlled – ideally at specified Entry Control Point (ECP)** – probably located at the inner cordon.

◆ all personnel who enter/exit the zone should be accounted for at the ECP

◆ all patients taken out of the zone should also be accounted for at the ECP (a log should be kept).

◆ All personnel within the Hot Zone should be in appropriate PPE.

◆ **ECP should be placed upwind from the source – at least 100m from the hazard** (greater distances may be required depending on the nature of the hazard. However, every effort should be made to ensure the ECP is outside of the radius of possible contamination.

◆ Vehicles may be used in the Hot Zone, but are then considered contaminated and should not cross into the Warm Zone (see below) until the incident is terminated. They must later be decontaminated.

WARM ZONE

The **Warm Zone is situated upwind and ideally uphill from the Hot Zone** (however, where geography and meteorology do not allow for this, the priority is for a location upwind). The Warm Zone is also likely to be within the inner cordon (because of its proximity vapour and cross contamination hazards). Activities in the Warm Zone may include:

◆ **Victims and first responders are decontaminated – probably through separate channels.**

◆ **Rescue, decontamination and medical personnel will be staged through the Warm Zone.**

◆ **All personnel must be in appropriate PPE.**

◆ **Movement between the inner and outer cordon must be logged**

◆ **Exit from the Warm Zone into the Cold Zone for all patients may be achieved most effectively via a separate '*patient transfer point*' / ambulance loading point .**

◆ **Rescue personnel working in the Warm Zone enter/exit the zone ONLY via the ECP rather than the patient transfer point / ambulance loading point** (to avoid congestion).

◆ **Typically, the Warm Zone will provide a 'buffer' at least 5 meters wide between the Hot and the Cold zones.** However, it but may need to be significantly wider, depending on the number of personnel working in the zone and the number of victims requiring decontamination.

Triage considerations. The rapid triage of victims is most likely to take place at the Warm Zone triage point.

(a) Patients are sent to immediate care area for life/limb-threatening injuries, delayed area for stable but non-ambulatory conditions, and ambulatory area for ambulatory patients.

(b) **Medical treatment is limited – sufficient only to stabilise the patients long enough to get through decontamination procedures.** This will usually involve only airway control, haemorrhage control and seizure control. Specific antidotes (e.g., atropine may be given to casualties caused by exposure to nerve agents).

(c) **Victims triaged into the 'immediate' category go through 'stretcher' decontamination first, followed by non-ambulatory delayed patients.**

(d) **Ambulatory victims go through ambulatory decontamination, as do all Hot Zone and Warm Zone personnel before entering the Cold Zone.**

(e) **After decontamination, all patients exit the Warm Zone via a patient transfer point / ambulance loading point.**

(f) No contaminated material, including stretchers, dressings and clothing should pass into the Cold Zone.

(g) Contaminated bodies and personal property recovered from the scene may be held in the Warm Zone

COLD ZONE*: This is upwind (and ideally uphill) from the Warm Zone and is also situated between the inner and outer cordons. The Cold Zone should be considered free of contamination but **all personnel must have PPE immediately to hand** (in case of a change in wind direction, the ineffective decontamination of personnel / materials from other zones, or a secondary release).

◆ **The purpose of this zone is to provide a safe environment for command and control vehicles, police investigation/evidence collection personnel, medical personnel and survivors.**

◆ Patients may enter via a patient transfer point / ambulance loading point and go through a cold triage point.

◆ Full spectrum of medical care may be done at pre-designated treatment areas, depending on medical supplies, personnel and expertise available.

◆ Separate arrangements may be in place for the movement of decontaminated bodies (for example, to a temporary mortuary); survivors (to a Survivor Rest Centre); casualties (through an ambulance loading point for onward movement to designated hospitals).

Cold Zone triage set-up procedures

Emergency medical considerations. Triage will begin after appropriately protected rescue personnel have released victims from the inner cordon. If an effective cordon is not in place, many walking victims are likely to leave the scene without being decontaminated and triaged and will go to medical facilities. **Decontamination and triage sites may therefore be necessary <u>at all nearby medical facilities</u>.**

Cold Zone triage: Set-up considerations

1.   Set-up site in direct line to the patient transfer point and consider:

◆   Location of roads

◆   Parking space for transport vehicles

◆   Location of decontamination team and vehicles (keep site separate from these).

Keep victims well away from rescue personnel to prevent chances of contamination.

---

2. Separate victims into four categories:

2.a. Immediate: Decontaminated victims who require immediate assistance because of complications of blast, agent or other.

◆ First Aid, including chemical agent antidotes, should be given by qualified personnel. Patients must be stabilised before transport.

Place victims on next available transport and direct to medical facility as instructed by Silver/ Bronze commander.

2.b. Delayed: Decontaminated victims who have effects that are not immediately life threatening and that can wait for definitive care (such as an uncomplicated arm fracture).

◆ Hold until Immediate victims are cared for. Send to medical facility with available space.

2.c. Minimal: Victims who can be treated on the scene, with subsequent release of the victim.

2.d. No effects: Victims who have not been contaminated and do not have effects.

◆ Hold for medical decision. These patients may be released at the discretion of the on-scene Medical Team Leader.

8.4 (D) Decontamination

The information in this Appendix is provided only as a general guide. First responders should be knowledgeable about the exact procedures used by their own organisation.

The details given below do not supersede any nationally agreed protocols or procedures.

Location of decontamination site. Ideally, decontamination should take place at a remote location within the Warm Zone. It should be upwind/uphill from the incident site and far enough away to be safe from any further blast or collapse. (Consideration should be given to using a location with facilities that may aid decontamination: car wash, swimming pools, fountains, indoor carparks with a sprinkler system, and so on.)

First principles. First responders should be familiar with two basic decontamination methods. These are chemical removal (or deactivation) and physical removal.

Chemical removal

Increasingly, the use of soap and water is the preferred option for removing chemical materials from personnel. However, **hypochlorite** and water (nine parts water, one part bleach) may be an appropriate option if its limitations are acceptable. Caution should be taken when using hypochlorite and water to decontaminate people because high concentrations will degrade the skin's ability to act as a barrier and increase the rate of absorption of a chemical agent. Under no circumstances should undiluted bleach be applied to exposed skin.

Diluted hypochlorite solutions. If hypochlorite is available, the 0.5 per cent solution can be made by adding 170 g (6 oz) of calcium hypochlorite to 19 litres (5 gallons) of water. Adding 8 times this amount (1.36 kg/48 oz) of calcium hypochlorite to 19 litres (5 gallons) of water will make a 5 per cent solution. The solutions should be made up fresh, with a pH in the alkaline range. After the solutions are prepared, they should be placed in marked buckets for use in the decontamination area.

Military perspective. Though the US military has changed its primary personnel decontamination solution to soap and water, hypochlorite and water remains a viable option. The US military uses two different concentrations of chlorine solution in patient decontamination procedures.

◆ A 0.5 per cent chlorine solution, which can be made using nine parts water and one part bleach, is a viable option for skin decontamination.

◆ **A 5 per cent chlorine solution (straight bleach)** is used by the military **for the decontamination of equipment** (for example, scissors, aprons and gloves used by personnel working in the patient decontamination area) and **should not be used on personnel**.

Note that bleach (hypochlorite) solutions should be rinsed off within 10 minutes of application.

Physical removal
Soap and water. Water and **water/soap solutions** can physically remove or dilute agents and **offer significant advantages for civilian use** because of safety, cost and availability considerations.

◆ Both fresh water and seawater have the ability to remove chemical agents, not only when forced through a hose, but also via slow hydrolysis. However, the generally low solubility and slow rate of diffusion of chemical materials in water significantly limits a material's hydrolysis rate.

Water or aqueous solutions . Copious flushing with water can often effectively remove an agent. This can be accomplished simply with the use of a fire hose and a spray nozzle. This method requires careful planning, however, because waste water will need to be safely collected. Cold water may also cause hypothermia in victims and rescue personnel.

Absorbent materials. Absorbent material can be used to reduce the amount of chemical agent that travels through the skin. In emergencies, dry powders such as soap detergents, earth and flour can be useful. The application of flour followed by wiping with wet tissue paper has been effectively used against various agents.

Mechanical removal. Scraping with a wooden stick (that is, a tongue depressor or 'lollipop' stick) can remove the bulk of an agent by physical means. One advantage of most physical removal methods is that they work equally well on all types of chemical agents regardless of composition

Field expedient decontamination
The following information may be appropriate in circumstances when rapid decontamination of a C or B material is deemed necessary, but specific decontamination equipment/specialists are not available.

Decontamination Solutions – summary
◆ 5 per cent hypochlorite is typically used for equipment, soap and water is used where skin contact may occur

◆ **Do not apply a 5 per cent hypochlorite solution directly to the skin**
◆ A serious increase in the likely absorption rate of chemical materials (and greatly enhanced skin damage with agents like *mustard*) will occur if a 5 per cent hypochlorite solution contacts the skin
◆ **Do not apply any hypochlorite mixtures to the face.** Hypochlorite in the eyes can cause cataracts.
◆ 0.5 per cent Hypochlorite solutions applied to the skin should be rinsed off within 10 minutes after application

Obtaining materials

Soaps and hypochlorite solutions can be found in any supermarket.

Soap. Liquid soaps, washing-up liquid and washing powders can be used.

Bleach. Most commercially available liquid bleaches (*Domestos*™, for example) state the percentage of hypochlorite solution (usually listed as 'less than 5%') on the container. At this percentage, hypochlorite solutions can be used on **equipment** directly from the bottle.

Calcium hypochlorite is also available in dry form as swimming pool/spa chlorine, and can be obtained from pool supply stores, building supply stores and hardware stores. (For example 50 lb of **granular chlorinizer** (65 per cent calcium hypochlorite) is sufficient to produce 80 gallons of 5 per cent hypochlorite solution, 800 gallons of 0.5 per cent solution)

Ancillary equipment. Plastic dustbins or similar plastic containers with 190-230 litre (50-60 gallon) capacity, along with sponges, brushes and pressurised garden sprayers can usually be obtained at the same locations.

Consideration should also be given to obtaining plastic sheeting, which can be used to control wastewater, or wrap 'dropped' equipment.

General decontamination procedures
Outer garments and equipment. Outer clothing is removed and the casualty is washed – without rubbing – with soap and water. This should be flushed off with copious amounts of water in a few minutes. Equipment is rinsed with 5 per cent hypochlorite solution or soap and water and allowed to dry.

Responders self/'buddy-buddy' decontamination
It is conceivable that emergency responders could become exposed to an agent during an incident. **It is vital that personnel do not panic if they suspect they have been exposed to an agent**. Survival and prevention of further exposure depends on the initial steps taken to physically remove an agent. The self/ buddy decontamination method is appropriate for first

responders and walking wounded (if physically capable).

◆ If sufficient resources are available, contaminated personnel and walking wounded should be placed in stations separated by gender.
◆ If sufficient resources are not available, decontamination should be performed according to the established standard.

Walking wounded (ambulatory) & self/'buddy-buddy' decontamination

1. **Remove** any signs of **gross contamination by scraping, sweeping** or **blotting** the material away.
2. **Remove clothing or equipment rapidly but cautiously**. Clothing should be pulled away from the body. In the event clothing needs to be removed over the head, cut it away.
3. After completing steps 1 and 2, **wash hands** before continuing the process.
4. **Remove all external** extraneous **items** from contact with the body. Such items include hearing aids, jewellery, watches, toupees, wigs and artificial limbs. (*Note:* if the victim or rescuer cannot safely evacuate the area without the use of eyeglasses, the glasses should be allowed to sit in a hypochlorite solution for at least 5 minutes.)

5. After removing eyeglasses or contact lenses, **flush the eyes with large amounts of water**.
6. Gently **wash face and hair with soap and lukewarm water**, followed by a thorough rinse with lukewarm water.
7. **Wash** body surfaces **with copious amounts of lukewarm soapy water**, rinsing in clear lukewarm water.
8. Change into uncontaminated clothing or blankets.

Decontamination of stretcher cases

If victims are unconscious, seriously injured or require extra care during decontamination, the most appropriate option may be to commence decontamination without removing them from the stretcher. The basic decontamination process for these patients is similar to self/buddy decontamination. The major difference is that more personnel and resources are required.

After decontamination, a victim should be monitored for possible contamination with a chemical agent monitor. Once a victim is confirmed clean of chemical agent, he or she should then be transferred through a 'shuffle pit' over the contamination control line. **A shuffle pit is composed of two parts super tropical bleach (STB) and three parts earth or sand. The pit should be deep enough to cover the bottom of the protective**

overboots. The decontamination apron and gloves should be washed in 5 per cent hypochlorite solution before the patient is transferred to a new clean stretcher or canvas stretcher. The patient then proceeds to the next station.

Stretcher patient decontamination

1. Remove any signs of gross contamination from victims before entering the decontamination station.
2. Transfer the patient to a decontamination preparation stretcher, cut away all clothing and remove all personal property. The patient should be transferred to a decontamination stretcher or a canvas stretcher with a plastic sheeting cover. All property should be bagged, secured and clearly identified.
3. **Eyeglasses and contact lenses**
    ◆ Rescuers' hands must be decontaminated by washing with copious amounts of soap and lukewarm water and then thoroughly rinsed with water before removing contact lenses.
    ◆ Contact lenses should be removed to decrease the risk of cross contamination.
    ◆ Contact lenses should be collected and discarded.

- ◆ Eyeglasses in metal frames can be decontaminated in a bath of hypochlorite solution for 5 minutes followed by a thorough rinsing.
- ◆ If eyeglasses are in a composite or plastic frame, they should be secured in an impermeable bag for later decontamination.

4. Decontamination team members should decontaminate their gloves and aprons (if they are wearing them) with the 5 per cent hypochlorite solution.

5. Wash victim with copious amounts of lukewarm soapy water. Superficial wounds are flushed and new dressings applied as needed. Splints are removed and treated as contaminated waste.

6. The victim should then be showered or washed thoroughly with copious amounts of water, starting with the face and hands and then the rest of the body.

7. Following the water decontamination, personnel should carry out medical screen procedures.

8. Following successful decontamination, an individual should have new (uncontaminated) dressings and splints applied to wounds. Patients should be transferred to the support area where fresh clothing is provided, and they are observed for further signs of exposure.

9. Each individual, having been processed through decontamination, should be marked and identified as such. This can be accomplished with a triage tag or by writing on a victim's forehead. During processing, each individual should receive a certificate indicating:
- Description of decontamination actions taken
- Time decontamination was completed
- Time released from observation area
- Any medical treatment performed in conjunction with decontamination.

(A copy should also go to decontamination record management.)

Transfer to medical facility
Casualties should be decontaminated before they are transported to a medical facility. (It is possible that the effect of off-gassing or liquid will endanger the ambulance personnel as ventilation is generally poor in ambulance compartments. In the Tokyo subway incident, a casualty exposed to vapour who had been removed from the subway tunnel by rescue personnel contaminated the ambulance staff who rode with him in the almost sealed ambulance.)

If decontamination is not possible before transporting a casualty, the ambulance personnel

must wear protective apparel, including a mask. If a contaminated patient is carried, the ambulance and ancillary equipment (such as stretchers and so on) may also become contaminated.

Under no circumstances should any covering be placed over or around the casualty (for example, blankets or 'space' blankets). Covering a contaminated casualty may increase both the rate of absorption and evaporation of chemical materials on their clothes or skin.

Disposal of contaminated clothing and equipment

It is likely that specific instructions will be issued by Gold or Silver command setting out the disposal arrangements for contaminated materials. However, if contaminated materials are accumulating and specific instructions have not been issued, the following procedure may be employed.

◆ Approximately 100 m downwind of the decontamination facility, an area should be designated as a clothing dump
◆ The dump should be clearly marked
◆ Contaminated equipment and clothing (excepting items for which decontamination facilities exist) should be placed in this dump
◆ If possible, the contaminated material should be placed in plastic bags, stored in closed airtight containers or covered with earth to prevent the escape of toxic vapours.

◆ The officer responsible for decontamination should always notify the incident commander of the existence of the dump and designate it on an area map.

Equipment decontamination procedures

In circumstances where contaminated equipment cannot be dumped, the following procedures, if authorised by the incident commander, may be appropriate.

◆ Nylon and canvas equipment bags may be decontaminated by boiling them for one hour in water. The addition of soap will speed this process against nearly all agents. After removal from the boiling water, rinse, air-dry and return the items to service. Such equipment can also be decontaminated by using bleach slurry methods.

◆ Leather equipment, such as belts, quickly absorbs liquid agents. Initial decontamination should be as rapid as possible. For a thorough decontamination, soak shoes, straps and other leather equipment for 4-6 hours in water heated to 50 to 55°C and then air dry without excess heat.

◆ Liquid contaminants found on impermeable protective clothing (including aprons, gloves and boots) should be neutralised or removed as quickly as possible. The quickest decontamination method can be performed while clothing is still being worn. If slurry is not available, blot liquid off with an available material, such as rags. This should be done immediately if splashes or large drops of agent contaminate clothing. However, a complete decontamination can be accomplished through one of the methods listed in the table below.

Equipment decontamination – options

Aeration	If the contamination is light or is caused by vapour, the articles can be decontaminated by airing them outdoors in the wind and sunlight for several days.
Water	◆ Immerse heavily contaminated articles in hot soapy water at a temperature just below boiling for one hour.
	◆ Do not stir or agitate. After one hour, remove the articles, rinse them in clear water and drain.
	◆ While items are still hot and wet, pull apart any surfaces that are stuck together and then hang them up to dry.
	◆ Repeat the process if necessary.
Slurry	Decontaminate impregnated items (worn by personnel) by spraying or applying slurry immediately after contamination. After a few minutes, wash off the slurry with water. This can be done while the clothing is being worn.

Care and decontamination of stretchers

Emergency protection for canvas or metal stretchers can be provided by covering them with materials such as military ponchos or plastic sheeting. All patient

Methods of stretcher decontamination

Canvas stretcher	Decontaminate canvas by immersion in boiling water for one hour. If available, add 4 lb of sodium carbonate (washing soda) to each 10 gallons of water. After boiling with washing soda, rinse with clear water.
Wood stretcher	Apply 30 per cent aqueous slurry of bleach and let it react for 12 to 24 hours. Repeat application if necessary, then swab the wood dry and let it aerate at elevated temperatures.
Metal stretcher	If the stretcher cannot be taken apart, decontaminate it with bleach slurry or by flushing it with hot soapy water. This method is effective for all chemical agents. Apply this solution to all contaminated surfaces by spray, broom or swab; after 30 minutes, flush with water. After decontamination, aerate the metal outdoors for several hours.
'Raven' stretcher	Soak in 5% hypochlorite solution for 30 minutes, then flush with water.

stretchers must be properly decontaminated to prevent further contact or vapour hazard.

Verifying decontamination

Despite the best efforts to completely decontaminate equipment, the possibility remains that a residual hazard may be present. Often, chemical agents are absorbed in porous materials. Such agents might later emerge as chemical vapours, posing a risk to both patients and medical personnel in a treatment area. Equipment should be checked with a chemical agent monitor (see Appendix A) before being placed into the general service or supply area.

If time allows, the following procedure provides an additional level of confidence that decontamination has been achieved.

(1) Individual items of equipment should be placed in separate clean plastic bags and the bags sealed.
(2) The bags should be placed in the sun or in a heated unoccupied structure and allowed to warm for 30 minutes.
(3) After 30 minutes, the bag should be slightly unsealed and the nozzle of the CAM should be placed into the opening to detect any indication of residual vapour. (*Note:* do not let the nozzle touch the bag itself and be careful not to 'swamp' the CAM)

(4) If residual contamination is found, the item should be disposed of in the appropriate manner as indicated for hazardous material waste, unless it is essential. If it is an essential piece of equipment, repeat the decontamination process and recheck.

(5) It is possible that at the scene of a large incident, several checking facilities will be necessary simultaneously.

Additional considerations

For chemical and biological agents, a warm soap and water wash with vigorous (non-abrasive) scrubbing of the affected parts will usually suffice to remove and hydrolyse the agent. Rapid decontamination is critical since most nerve agents can cause injury in several minutes and removal of the biological materials will prevent further spread to other equipment due to secondary aerosols.

◆ An expeditious manner in which to conduct emergency decontamination operations is to use small commercial paddling/swimming pools. In addition, the use of commercial swimming pool portable chlorine test kits can ensure the proper mixture of water chlorine.

◆ Care should be taken when using bleach in an enclosed space to prevent off-gassing of chlorine gas.

◆ If an agent is released in a building in which there are people, either the agent or the people must be removed from the building as quickly as possible.

(To remove the agent, the intake ducts and vents should be closed – in the event the agent was in the air supply – and the exhaust system should be maximised to allow the egress of air as quickly as possible. It may not be feasible to exhaust the air (containing the agent) into the outside air. The alternative is to move people outside as quickly as possible, to minimise exposure time, and to decontaminate. Consult the incident commander.)

◆ Check with local environmental officials concerning run-off and containment considerations.

◆ PVC (plastic) pipe becomes brittle in cold temperatures and the glue used to hold the PVC together becomes much less adhesive. All PVC piping must be inspected and tested before use.

◆ Additional arrangements may need to be made for the decontamination and storage of contaminated cadavers.

◆ Chain of custody and preservation of evidence must be established.

◆ Decontamination teams should be established for all hospitals to prevent the spread of contamination.

◆ Always be aware of hydrant pressures and water availability.

◆ Consider night decontamination operations.

◆ Consider animal decontamination (guide dogs, drug-sniffing dogs, and so on)

◆ Medicants and doses of antidotes for adults and children vary greatly and should be considered during pre-planning.

8.5 (E) Phased training guide

Guidance notes for officers with training responsibilities

Training is essential to a rapid, concerted, and safe response. In devising the training requirement, the following should be considered:

◆ **Make a training/rehearsal schedule**
◆ **Ensure the training is relevant**
◆ **Practise** your response plan regularly (according to the schedule), as response skills are 'perishable'
◆ **Incorporate the use of equipment** as much as possible in your training plan
◆ **Incorporate the use of a call list** during the training process
◆ **Incorporate a multi-agency dimension**
◆ **Encourage responders to rehearse and drill.** Responders should be required to employ the skills they are trained in, perform the tasks they would be expected to accomplish in a given situation and successfully complete these tasks under conditions that simulate an emergency response environment.
◆ A training plan can be organised in the follow sequence of phases:
  (a) Crawl – Talk through the response plan indicating key points and having responders perform the functions as you explain them.
  (b) Walk – Walk through the response plan explaining and supervising the process as

responders perform their duties at half speed. During a walk-through responders can still ask questions of the instructors. In addition, during this phase instructors should ask questions, have responders explain what they are doing and raise the issue of possible contingencies or complications that may occur.

(c) Run – Perform the response plan as you would if the event was actually occurring. Responders should move with a sense of urgency during this phase. In addition, training should create a sense of uncertainty in the situation, creating scenarios that challenge responders to adapt to and overcome unexpected complications.

(d) Review – Perform an *after-action* review of the training you have just initiated. Ask the responders what went right, what went wrong, and how they plan to improve. Encourage all responders to provide feedback. After the responders have commented, the trainer should provide feedback based on what went right, what went wrong, and how the response could improve.

◆ When responders have completed training, the appropriate details should be kept with their personal records or within some centralised training record

8.6 (F) Case studies

8.6.1 Chemical case one

A 40-year-old man entered the casualty department because of sudden onset of dim vision, rhinorrhoea and slight respiratory discomfort. He had been working in a room where large containers of sarin were stored. Although there were other workers in the room he was the only one in close proximity to the containers. He entered the casualty department about 20 minutes after the onset of symptoms.

He reported that suddenly his vision became dim – "like I had just put on sunglasses" – his nose started running and he became moderately short of breath. He further reported that his breathing had improved considerably since the onset.

On examination, his pupils were about 1.5 mm bilaterally, there was moderate conjunctival injection, he had slight rhinorrhoea and very mild respiratory distress. No antidotes were administered.

Over the next several hours, both the rhinorrhoea and breathing improved and at time of discharge, four hours after he entered the emergency room, he had only slight irritation in his eyes and dim vision.

His eyes were examined in the dark at frequent intervals over the next several months. They did not regain their ability to dilate fully in the dark for seven weeks after exposure.

Comments

This is a typical casualty from exposure to a small amount of sarin vapour: eye, nose and airway effects.

His breathing and rhinorrhoea improved considerably during the 20 minutes after exposure. This often occurs following small or moderate exposures.

Because his breathing caused him no discomfort, he was given no antidotes.

The pupils did not return to normal until seven weeks post-exposure. However, when examined in moderate (room) lighting they appeared to be normal after the first week.

8.6.2 Chemical case two

A 52-year-old man entered the casualty department convulsing and cyanotic, with laboured breathing, muscular fasciculations, miosis, and marked rhinorrhoea and salivation.

He had been working in a sarin-contaminated area. Although he had been in full protective gear, it was later found that there was a small leak in the voicemitter diaphragm of his mask. He noted an increase in oral and nasal secretions and difficulty breathing. Within minutes he was in marked respiratory distress and convulsing.

He was immediately given 4 mg of atropine, an infusion of 2-PAMCl was begun and oxygen was given. Cyanosis, secretions and bronchoconstriction diminished. However, about 20 minutes later secretions increased and more atropine was given and another infusion of 2-PAMCl was begun. He briefly became semiconscious, but a few minutes later relapsed and became cyanotic, comatose and apnoeic.

He required ventilation with oxygen and repeated doses of atropine for increasing bronchoconstriction over the next several hours. About 2.5 hours after admission his sensorium slowly began to clear and he began to breathe spontaneously. At 9 hours after admission he walked around the room although he was weak and areflexic.

Comments

This patient inhaled sarin vapour that leaked through the voicemitter on his protective mask. The absorption of the agent was apparently slow because of the slow progression of effects.

Although the effects were severe, he responded to repeated doses of atropine.

Ventilation was required for about 2.5 hours.

8.6.3 Chemical case three

This patient was one of several people who were investigating the leak of about 100 gallons of sarin in a field. Ignoring advice, he refused to wear a mask. The group left vehicles about 183 m upwind of the puddle, and the patient then left the group to be the first to get close to the puddle. He walked to within 3 m of the puddle, and then was seen to turn, clutch his chest and collapse unconscious. By the time the others arrived he was convulsing. In 3 to 4 min his convulsions stopped because he was apnoeic and flaccid.

He was immediately given atropine (no 2-PAMCl was available), and ventilation was begun as soon as the others could carry him back to the vehicles. He required repeated doses of atropine and assisted ventilation over the following 3 hours, when he started regaining consciousness and muscular control. He went on to a complete recovery.

Comments

This individual inhaled a large concentration of sarin in apparently one or two breaths.

The first effect was loss of consciousness, which occurred within seconds after he inhaled the agent. He then convulsed, and the convulsions were halted by apnoea and flaccid paralysis.

Without the almost immediate administration of atropine and ventilation support with oxygen this patient probably would have died within a short period.

8.6.4 Biological case one

In September-November 2001, 22 cases of anthrax infection occurred in locations across the eastern seaboard of the United States. Eleven cases were confirmed as inhalation anthrax, while the remaining 11 (seven confirmed, four suspected) were cutaneous. Of the inhalation cases, six patients (54.5 per cent) survived. All cutaneous patients recovered with antibiotic therapy. Exposure to anthrax-contaminated items sent via the US Postal Service was shown to be the source of infection in nine of the 11 inhalation cases. The source of exposure in the remaining two inhalation cases was never determined.

Comments

There are three clinical forms of anthrax: cutaneous, gastrointestinal and inhalation. First responders and medical personnel should be aware that the early symptoms of inhalation anthrax, the most likely form of anthrax to be seen in an intentional release, are non-specific and flu-like. A chest X-ray or CT scan is considered to be a sensitive early indicator of infection, as radiological evidence of a widened mediastinum is pathognomic for inhalation anthrax. Early diagnosis is essential for recovery, as early antibiotic therapy (with aggressive supportive therapy) is linked to increased survival (previous reports put the survival rate of inhalation anthrax at <15 per cent) and antibiotic therapy has little impact when initiated late in infection. A later presumptive case showed many clinical aspects of inhalation *B. anthracis* infection, including widened mediastinum, but with negative blood cultures. This case was linked to an interrupted course of post-exposure prophylactic therapy and highlights both the difficulties in case confirmation and the importance of completion of the prescribed prophylactic.

8.6.5 Biological case two

In 1979 at a city in the former Soviet Union, Sverdlovsk, an unusual anthrax outbreak occurred. Soviet officials attributed it to consumption of contaminated meat. However, officials in the US attributed it to a release from a military microbiology facility in the city. Sixty-eight people died and an undetermined number were made sick. In spite of massive antibiotic therapy, few of the affected survived.

Comments

First responders need to review the epidemiology of the outbreak and note the deviations from normal outbreaks, such as number of people affected, time relationship, clinical picture, geographical boundaries and other species affected. In this case, epidemiological data show that most victims lived in a narrow zone extending from the military facility to the southern city limit. Farther south, livestock also died of anthrax along the zone's extended axis. The affected zone paralleled the northerly wind that prevailed shortly before the outbreak. It is concluded that the escape of an aerosol of anthrax at the military facility caused the outbreak.

8.6.6 Biological case three

Based on the extensive vulnerability studies performed in the 1950s and 1960s, several theoretical but highly likely terrorist scenarios can be developed. A biological agent, *Francisella tularensis* (tularemia), a highly infectious incapacitating agent with an infective dose of 10 to 100 cells, could be easily produced in a clandestine location by a graduate student using readily available medium ingredients on about 100 blood agar petri dishes. This would yield 10 g of dry material, which could be filled into as few as six 100-watt light bulbs. These bulbs would be deposited covertly on the subway tracks in any large city although this test was initially performed in New York City. The NYC north-south system contains about 4.75×10^8 litres of air. Riders on the subway for at least 5 min would receive greater than 10 human doses if breathing at the rate of 10 litres per minute. Secondary aerosols would be generated each time a train crossed over the impact area. At least several hundred thousand subway riders would be affected with varying onset times linked to the dose received. This could be likened to the Tokyo subway attack with sarin nerve agent.

Comments

First responders should be aware of the potential for their own contamination, especially after observation of the initial casualties. Measures for decontamination of the personnel affected, the subway cars and tracks and any medical facilities where casualties were evacuated should be undertaken. At least a warm soap-and-water wash is recommended in any case. Following the examination of casualties, epidemiological studies to differentiate natural versus deliberate causes could be initiated.

8.7 (F) Glossary of medical terms

Adapted from *Webster's Third New International Dictionary* and *Stedman's Electronic Medical Dictionary*.

acidosis
A condition of decreased akalinity of the tissues of the blood.

active immunisation
The production of active immunity.

adenopathy
Swelling or morbid enlargement of the lymph nodes.

---

analgesic
1. A compound capable of producing analgesia, that is, one that relieves pain by altering perception of nociceptive stimuli without producing anaesthesia or loss of consciousness.

2. Characterized by reduced response to painful stimuli.

anticonvulsant
An agent which prevents or arrests seizures.

apnoea
Temporary cessation of respiration.

asthenia
Weakness or debility.

atropine
An anticholinergic, with diverse effects (tachycardia, mydriasis, cycloplegia, constipation, urinary retention) attributable to reversible competitive blockade of acetylcholine at muscarinic-type cholinergic receptors; used in the treatment of poisoning with organophosphate insecticides or nerve gases.

auscultation
The act of listening to the heart, lungs or other organs with a stethoscope or by direct application of the ear to the body.

Brucella
A genus of encapsulated, non-motile bacteria (family Brucellaceae) containing short, rod-shaped to coccoid, gram-negative cells. These organisms do not produce gas from carbohydrates, are parasitic, invading all animal tissues and causing infection of the genital organs, the mammary gland, and the respiratory and intestinal tracts. They are pathogenic for man as well as for various species of domestic animals.

bubo
Inflammatory swelling of one or more lymph nodes, usually in the groin; the confluent mass of nodes usually suppurates and drains pus.

bulla, gen. and pl. bullae
A large blister appearing as a circumscribed area of separation of the epidermis from the sub-epidermal structure (sub-epidermal bulla) or as a circumscribed area of separation of epidermal cells (intra-epidermal bulla) caused by the presence of serum, or occasionally by an injected substance.

CANA
Abbreviation for convulsive antidote, nerve agent.

coccobacillus
A short, thick bacterial rod of the shape of an oval or slightly elongated coccus.

conjunctivitis
Inflammation of the mucous membrane that lines the inner surface of the eyelids and is continued over the forepart of the eyeball covering part of the sclerotic coat and forming epithelium over the cornea.

cutaneous
Relating to the skin.

cyanosis
A dark bluish or purplish coloration of the skin and mucous membrane due to deficient oxygenation of the blood, evident when reduced haemoglobin in the blood exceeds 5 g/100 ml.

distal
Situated away from the centre of the body, or from the point of origin; specifically applied to the extremity or distant part of a limb or organ.

DS2
Abbreviation for decontamination solution 2.

dyspnoea
Shortness of breath, a subjective difficulty or distress in breathing, usually associated with disease of the heart or lungs; occurs normally during intense physical exertion or at high altitude.

---

enanthem, enanthema
A mucous membrane eruption, especially one occurring in connection with one of the exanthemas.

endotoxaemia
Presence in the blood of endotoxins.

enterotoxin
A cytotoxin specific to the cells of the intestinal mucosa.

erythema
Redness of the skin due to capillary dilation.

exanthema
A skin eruption occurring as a symptom of an acute viral or coccal disease, as in scarlet fever or measles.

fasciculation
Involuntary contractions, or twitchings, of groups (fasciculi) of muscle fibres, a coarser form of muscular contraction than fibrillation.

febrile
Denoting or relating to fever.

Glanders
A chronic debilitating disease of horses and other equids, as well as some members of the cat family, caused by *Pseudomonas mallei*; it is transmissible to humans. It attacks the mucous membranes of a horse's

nostrils, producing an increased and vitiated secretion and discharge of mucus, and enlargement and induration of the glands of the lower jaw.

immunoassay
Detection and assay of substances by serological (immunological) methods; in most applications the substance in question serves as antigen, both in antibody production and in measurement of antibody by the test substance.

IND
Investigational New Drug.

inoculation
Introduction into the body of the causative organism of a disease.

in vitro
In an artificial environment, referring to a process or reaction occurring therein, as in a test tube or culture media.

in vivo
In the living body, referring to a process or reaction occurring therein.

macula, pl. maculae
1. A small spot, perceptibly different in colour from the surrounding tissue.

2. A small, discoloured patch or spot on the skin, neither elevated above nor depressed below the skin's surface.

miosis
Excessive smallness or contraction of the eye.

mydriatic
Causing or involving dilation of the pupil.

oedema
An accumulation of an excessive amount of watery fluid in cells, tissues, or serous cavities.

PCR
Abbreviation for polymerise chain reaction.

phosgene
Carbonyl chloride; a colourless liquid below 8.2°C, an extremely poisonous gas at ordinary temperatures; it is an insidious gas, since it is not immediately irritating, even when fatal concentrations are inhaled.

pulmonary oedema
Oedema of the lungs.

radiomimetic
Producing effects (especially biological) similar to those produced by radiation.

rhinorrhoea
A discharge from the nasal mucous membrane.

sarin
A nerve poison which is a very potent irreversible cholinesterase inhibitor and a more toxic nerve gas than tabun or soman.

soman
An extremely potent cholinesterase inhibitor.

zoonosis
An infection or infestation shared in nature by humans and other animals that are the normal or usual host; a disease of humans acquired from an animal source.

8.8 (H) Precursor chemicals

[a] Figures in parentheses are based on the use of PCl3 as a chlorine donator in the reaction.

[b] Thionyl chloride could serve as chlorinating agent in all of these processes – other chlorinating agents could be substituted.

2-Chloroethanol 107-07-3

Civil uses	CW agent production	Agent units/precursor unit[a]
Organic synthesis	Sulfur mustard (HD)	0.99
Manufacturing of ethylene-oxide and ethylene-glycol	Sesqui mustard	0.99
	Nitrogen mustard (HN-1)	1.06
Insecticides		
Solvent		

3-Hydroxy-1-methylpiperidine 3554-74-3

Civil uses	CW agent production	Agent units/precursor unit[a]
Specific uses not identified. Probably used in pharmaceutical industry.	Non-identified. Could probably be used in the synthesis of psychoactive compounds such as BZ.	

3-Quinuclidinol 1619-34-7

Civil uses	CW agent production	Agent units/ precursor unit[a]
Hypotensive agent	BZ	2.65
Probably used in synthesis of pharmaceuticals		

3-Quinuclidone 1619-34-7

Civil uses	CW agent production	Agent units/ precursor unit[a]
Same as 3-quinuclidinol	BZ	2.65
3-quinuclidinol		

Ammonium bifluoride 1341-49-7

Civil uses	CW agent production	Agent units/ precursor unit[a]
Ceramics	Sarin (GB)	2.46
Disinfectant for food equipment	Soman (GD)	3.20
Electroplating	GF	3.16
Etching glass		

Arsenic trichloride 7784-34-1

Civil uses	CW agent production	Agent units/ precursor unit[a]
Organic synthesis	Arsine	0.43
Pharmaceuticals	Lewisite	1.14
Insecticides		
Ceramics	Adamsite (DM)	1.53
	Diphenylchloroarsine (DA)	1.45

Benzilic acid 76-93-7

Civil uses	CW agent production	Agent units/ precursor unit[a]
Organic synthesis	BZ	1.48

Diethyl ethylphosphonate

Civil uses	CW agent production	Agent units/ precursor unit[a]
Heavy metal extraction	Ethyl sarin (GE)	0.93
Gasoline additive		
Antifoam agent		
Plasticizer		

Diethyl methylphosphonite 15715-41-0

Civil uses	CW agent production	Agent units/ precursor unit[a]
Organic synthesis	VX	1.97

Diethyl N,N-dimethyl phosphoramidate 2404-03-7

Civil uses	CW agent production	Agent units/ precursor unit[a]
Organic synthesis	Tabun (GA)	0.90
Specific uses not identified		

Diethylaminoethanol 100-37-8

Civil uses	CW agent production	Agent units/ precursor unit[a]
Organic synthesis	VG	2.30
Anticorrosion compositions	VM	2.05
Pharmaceuticals		
Textile softeners		

Diethylphosphite 762-59-2

Civil uses	CW agent production	Agent units/ precursor unit[a]
Organic synthesis	VG	Catalyst
Paint solvent	Sarin (GB)	1.02
Lubricant additive	Soman (GD)	1.32
	GF	1.30

Diisopropylamine 108-18-9

Civil uses	CW agent production	Agent units/ precursor unit[a]
Organic synthesis	VX	3.65
Specific uses not identified		

Dimethyl ethylphosphonate 6163-75-3

Civil uses	CW agent production	Agent units/ precursor unit[a]
Organic synthesis	Ethyl sarin (GE)	1.12

Dimethyl methylphosphonate (DMMP) 756-79-6

Civil uses	CW agent production	Agent units/ precursor unit[a]
Flame retardants	Sarin (GB)	1.12
	Soman (GD)	
	GF	1.45

Dimethylamine 124-40-3

Civil uses	CW agent production	Agent units/ precursor unit[a]
Organic synthesis	Tabun (GA)	3.61
Pharmaceuticals		
Detergents		
Pesticides		
Gasoline additive		
Missile fuels		
Vulcanization of rubber		

Dimethylamine HCl 506-59-2

Civil uses	CW agent production	Agent units/ precursor unit[a]
Organic synthesis	Tabun (GA)	1.99
Pharmaceuticals		
Surfactants		
Pesticides		
Gasoline additives		

Dimethylphosphite 868-85-9

Civil uses	CW agent production	Agent units/ precursor unit[a]
Organic synthesis	Sarin	1.27
Lubricant additive	Soman	1.65
	GF	1.65

Ethylphosphonous dichloride 1498-40-4

Civil uses	CW agent production	Agent units/ precursor unit[a]
Organic synthesis	VE	1.93
Specific uses not identified but could be used in manufacturing of flame retardants, gas additives, pesticides, surfactants and so on	VS	2.14
	Ethyl sarin (GE)	1.18

Ethylphosphonous difluoride 430-78-4

Civil uses	CW agent production	Agent units/ precursor unit[a]
Organic synthesis	VE	2.58
	Ethyl sarin (GE)	1.57

Ethylphosphonyl dichloride 1066-50-8

Civil uses	CW agent production	Agent units/ precursor unit[a]
Organic synthesis	Ethyl sarin (GE)	2.10
Specific uses not identified. See ethylphosphonous dichloride.		

Ethylphosphonyl difluoride 753-98-0

Civil uses	CW agent production	Agent units/ precursor unit[a]
Organic synthesis	Ethyl sarin (GE)	2.70
Specific uses not identified. See ethylphosphonous dichloride.		

Hydrogen fluoride

Civil uses	CW agent production	Agent units/ precursor unit[a]
Fluorinating agent in chemical reactions	Sarin (GB)	7.0
Catalyst in alkylation and polymerization reactions	Soman (GD)	9.11
Additives to liquid rocket fuels	Ethyl sarin (GE)	7.7
Uranium refining	GF	9.01

Methyl benzilate 76-89-1

Civil uses	CW agent production	Agent units/ precursor unit[a]
Organic synthesis	BZ	1.39
Tranquilizers		

Methylphosphonous dichloride 676-83-5

Civil uses	CW agent production	Agent units/ precursor unit[a]
Organic synthesis	VX	2.28

Methylphosphonous difluoride 753-59-3

Civil uses	CW agent production	Agent units/ precursor unit[a]
Organic Synthesis	VX	3.18
	VM	2.84
	Sarin (GB)	1.67
	Soman (GD)	2.17
	GF	2.15

Methylphosphonyl dichloride 676-97-1

Civil uses	CW agent production	Agent units/ precursor unit[a]
Organic synthesis	Sarin (GB)	1.05
Specific uses not identified	Soman (GD)	1.36
	GF	1.35

Methylphosphonyl difluoride 676-99-3

Civil uses	CW agent production	Agent units/ precursor unit[a]
Organic syntheses	Sarin (GB)	1.40
Specific uses not identified	Soman (GD)	1.82
	GF	1.80

N,N-diisopropyl-2-aminoethyl chloride hydrochloride

Civil uses	CW agent production	Agent units/ precursor unit[a]
Organic synthesis	VX	1.34

N, N-diisopropyl-aminoethanethiol 5842-07-9

Civil uses	CW agent production	Agent units/ precursor unit[a]
Organic synthesis	VX	1.66
	VS	1.75

N,N-diisopropyl-(beta)-aminoethanol 96-80-0

Civil uses	CW agent production	Agent units/ precursor unit[a]
Organic synthesis	VX	1.84
Specific uses not identified		

N, N-diisopropyl-(beta)-aminoethyl chloride 96-79-7

Civil uses	CW agent production	Agent units/ precursor unit[a]
Organic synthesis	VX	1.64
	VS	1.72

**O-ethyl,2-diisopropyl aminoethyl methyl-phosphonate (QL) 57856-11-8

Civil uses	CW agent production	Agent units/ precursor unit[a]
Specific uses not identified	VX	1.14

Phosphorus oxychloride 10025-87-3

Civil uses	CW agent production	Agent units/ precursor unit[a]
Organic synthesis	Tabun (GA)	1.05
Plasticizers		
Gasoline additives		
Hydraulic fluids		
Insecticides		
Dopant for semiconductor grade silicon		
Flame retardants		

Phosphorous pentachloride 10026-13-8

Civil uses	CW agent production	Agent units/ precursor unit[a]
Organic synthesis	Tabun (GA)	0.78
Pesiticides		
Plastics		

Phosphorus pentasulfide 1314-80-3

Civil uses	CW agent production	Agent units/ precursor unit[a]
Organic synthesis	VG	1.21
Insecticide	VX	1.20
Mitocides		
Lubricant oil additives		
Pyrotechnics		

Phosphorus trichloride 7719-12-2

pugCivil uses	CW agent production	Agent units/ precursor unit[a]
Organic synthesis	VG	1.95
Insecticides	Tabun (GA)	1.18
Gasoline additives	Sarin (GB)	1.02
	Salt process	(0.34)
Plasticizers	Rearrangement process	1.02
		(0.68)
Surfactants	Soman (GD)	1.32
Salt process		(0.44)
Dyestuffs	Rearrangement process	1.32
		(0.88)
	GF	1.31
	Salt process	(0.44)
	Rearrangement process	1.31
		(0.87)

Pinacolone 75-97-8

Civil uses	CW agent production	Agent units/ precursor unit[a]
Specific uses not identified	Soman (GD)	1.82

Pinacolyl alcohol 464-07-3

Civil uses	CW agent production	Agent units/ precursor unit[a]
Specific uses not identified	Soman (GD)	1.79

Potassium bifluoride 7789-29-9

Civil uses	CW agent production	Agent units/ precursor unit[a]
Fluorine production	Sarin (GB)	1.79
Catalyst in alkylation	Soman (GD)	2.33
Treatment of coal to reduce slag formation	GF	2.31
Fluid in silver solder		

Potassium cyanide 151-50-8

Civil uses	CW agent production	Agent units/ precursor unit[a]
Extraction of gold and silver from ores	Tabun (GA)	1.25
Pesticide		
Fumigant	Hydrogen cyanide	0.41
Electroplating		

Potassium fluoride 7789-23-3

Civil uses	CW agent production	Agent units/ precursor unit[a]
Fluorination of organic compounds	Sarin (GB)	2.41
Cleaning and disinfecting brewery, dairy and other food processing equipment	Soman (GD)	3.14
Glass and porcelain manufacturing	GF	3.10

Sodium bifluoride 1333-83-1

Civil uses	CW agent production	Agent units/ precursor unit[a]
Antiseptic	Sarin (GB)	2.26
Neutralizer in laundry operations	Soman (GD)	2.94
Tin plate production	GF	2.91

Sodium cyanide 143-33-9

Civil uses	CW agent production	Agent units/ precursor unit[a]
Extraction of gold and silver from ores	Tabun (GA)	1.65
Fumigant	Hydrogen cyanide	0.55
Manufacturing dyes and pigments	Cyanogen chloride	1.25
Core hardening of metals		
Nylon production		

Sodium fluoride 7681-49-4

Civil uses	CW agent production	Agent units/ precursor unit[a]
Pesticide	Sarin (GB)	3.33
Disinfectant	Soman (GD)	4.34
Dental prophylaxis	GF	4.29
Glass and steel manufacturing		

Sodium sulfide 1313-82-2

Civil uses	CW agent production	Agent units/ precursor unit[a]
Paper manufacturing	Sulphur mustard (HD)	2.04
Rubber manufacturing		
Metal refining		
Dye manufacturing		

Sulphur dichloride 10545-99-0

Civil uses	CW agent production	Agent units/ precursor unit[a]
Organic synthesis	Sulphur mustard (HD)	1.54
Rubber vulcanizing		
Insecticides		
Vulcanizing oils		
Chlorinating agent		

Sulphur monochloride sulfur chloride 10025-67-9

Civil uses	CW agent production	Agent units/ precursor unit[a]
Organic synthesis	Sulphur mustard (HD)	1.18
Pharmaceuticals		
Sulfur dyes		
Insecticides		
Rubber vulcanization		
Polymerization catalyst		
Hardening of soft woods		
Extraction of gold from ores		

Thiodiglycol 111-48-8

Civil uses	CW agent production	Agent units/ precursor unit[a]
Organic synthesis	Sulphur mustard (HD)	1.3
Carrier for dyes in textile industry		
Lubricant additives	Sesqui mustard (Q)	1.79
Manufacturing plastics		

Thionyl chloride[b] 7719-09-7

Civil uses	CW agent production	Agent units/ precursor unit[a]
Organic Synthesis	Sarin (GB)	1.18
	Soman (GD)	1.53
	GF	1.51
	Sulphur mustard (HD)	1.34
Chlorinating agent	Sesqui mustard (Q)	1.84
Catalyst	Nitrogen mustard (HN-1)	0.714
Pesticide	Nitrogen mustard (HN-2)	0.655
Engineering plastics	Nitrogen mustard (HN-3)	1.145

CHAPTER 8: Appendix

Triethanolamine 102-71-6

Civil uses	CW agent production	Agent units/ precursor unit[a]
Organic synthesis	Nitrogen mustard (HN-3)	1.37
Detergents		
Cosmetics		
Corrosion inhibitor		
Plasticiser		
Rubber accelerator		

Triethanolamine hydrochloride (HN-3)

Civil uses	CW agent production	Agent units/ precursor unit[a]
Organic synthesis	Nitrogen mustard	1.10
Insecticides		
Surface active agents		
Waxes, polishes		
Textile specialities		
Lubricants		
Toiletries		
Cement additive		
Petroleum demulsifier		
Synthetic resin		

Triethyl phosphite 122-52-1

Civil uses	CW agent production	Agent units/ precursor unit[a]
Organic synthesis	VG	1.62
Plasticisers		
Lubricant additives		

Trimethyl phosphite 121-45-9

Civil uses	CW agent production	Agent Units/ Precursor Unit[a]
Organic synthesis	Used to make dimethyl methyl-phosphonate (DMMP)-molecular rearrangement	See (DMMP)

INDEX

A

AC 37, 72, 73

Aerosol 11, 40, 46, 47, 52-55, 97, 130-134, 136-138, 140-143, 148, 149, 152, 155, 157, 159, 161, 164, 166, 168, 170, 173, 176, 178, 180-182, 201, 209, 215, 274
  BW delivery 11, 130-134, 138, 140, 141, 148

Anthrax 5, 130, 138, 143, 146-148, 187, 188, 240
  effects 147
  treatment 146

Antibiotics 116-118, 147, 152, 187, 190, 194, 195, 200, 211, 214 217

Antidote 105, 106, 112, 115, 116, 118, 123, 254
  2-PAMCl 106, 108, 110

Atropine 105, 106, 108-110, 205, 254
  potential negative effects 106, 109, 205

B

Bacillus anthracis 141, 146

Bacteria 134, 135, 146-157, 187-200
  effects 146, 158
  treatment 146, 187-200

Bentonite 138

Biological agent quick reference 130-133

Biological detection 11, 239

Bleach 29, 121, 259, 260, 262, 274

G

H

I

Impure Sulphur Mustard 81, 82
Incubation period 130-134, 146, 147, 149, 152, 153,156, 157, 159-161, 172, 176-179, 182, 183, 184
Infectivity 134, 159
Ingestion 148, 152, 155, 162, 163, 165-170
Inhalation 44, 47, 54, 70, 86, 93, 97, 113, 116, 130, 142 147, 148, 152, 154, 155, 160-163, 165, 168-170, 182, 187, 205

J

Junin virus 177
   effects 177

L

L 38, 85, 86
Legionnaires disease 161
Lewisite 43, 46, 80, 85, 86, 89, 90, 94, 115; *see also Blister Agents*
   effects 80, 89, 90, 94
   treatment 115
Line-source; *see Delivery systems (biological)*
Liquid mustard 54, 56
Lorazepam 109
Litter patients 120, 121, 254, 265, 266; *see also Decontamination*
Lung agents; *see Pulmonary agents*

M

N

O

P

S

T

V

CHALET Standard Situation Report Format

C   Casualties – number by triage category, trapped, type of injury

H   Hazards – actual or potential

A   Access – the best access to a site for ambulances

L   Location – exact location for attending resources including aircraft

E   Emergency Services – those present and those required

T   Type – type of incident and what exactly is involved